Would you like to discover a comet? Or maybe you would like to be the first person on Earth — perhaps in the Universe — to recognise a celestial newcomer such as a nova or supernova. This *Guide* tells you how to become an astronomical discoverer using only modest equipment. It describes what approaches to take in searching the night skies for the unusual and the unexpected. Included are 20 contributions from both amateur and professional astronomers who are world famous for their discoveries.

This book covers all the kinds of objects that an amateur can hope to find as a result of systematic searching. Personal contributions from noted discoverers such as David Levy and Minoru Honda (comets), Bob Evans (supernovae), Eleanor Helin and Brian Manning (asteroids), and Mike Collins and Dan Kaiser (variable stars) reveal the secrets of their successful methods.

Visual, photographic, and electronic search techniques are described, as well as instructions on how to report discoveries. Extensive appendices pack in a wealth of useful data for every new discoverer of cosmic phenomena.

Dr Bill Liller, a professional astronomer for 30 years at Harvard University and the University of Michigan, took early retirement and began a second astronomical career — as an amateur astronomer. Bill has discovered more novae and nova-like objects than anybody else in history (16 as of 1992), and is still going strong. He has also found a comet (1988V), two supernovae in remote galaxies, and the asteroid (3040) Kozai. Bill currently works part-time for the Instituto Isaac Newton in Santiago, Chile.

The Cambridge guide to astronomical discovery

THE CAMBRIDGE GUIDE TO ASTRONOMICAL DISCOVERY

William Liller

CAMBRIDGE
UNIVERSITY PRESS

Published by the Press Syndicate of the University of Cambridge
The Pitt Building, Trumpington Street, Cambridge CB2 1RP
40 West 20th Street, New York, NY 10011-4211, USA
10 Stamford Road, Oakleigh, Victoria 3166, Australia

First published 1992

Printed in the United States of America

A catalogue record for this book is available from the British Library

Library of Congress cataloguing in publication data

Liller, William, 1927–
The Cambridge guide to astronomical discovery / William Liller.
p.   cm.
Includes index.
ISBN 0-521-41839-9
1. Astronomy–Observers' manuals.   2. Astronomy–Amateurs'
manuals.   I. Title.
QB64.L55   1992
520–dc20   92-2705-CIP

ISBN 0 521 41839 9 hardback

Contents

Preface

This book is not for everyone. It has been written exclusively for those who warm to excitement with the news of a new astronomical discovery – a comet that just might decorate the evening sky with a long sweeping tail, an asteroid that, according to first predictions, will tumble by the Earth at a distance of less than a mere million kilometres, or a nova or supernova that could possibly outshine all other stars in the sky.

Perhaps you, the reader, just do not have the time to join in the hunt and simply want to read about it, to become vicariously involved. Fine; the bulk of the book should be just what you want. Or maybe you desire to join in the fray and would like to learn how to get started. Also, fine; there are several chapters and an extensive Appendix just for you. Or, possibly, you are already one of the hunters and would like to improve your chances of turning up a new visitor to the skies. Good, because one chapter has been written by the more successful searchers of the sky and includes numerous useful tips and invaluable suggestions. Also, the tables in Appendix F, looking much like the financial barometer pages of a big city newspaper, chart in detail the circumstances of those discoveries made during the decade of the 1980s.

Let it be known that *anybody* can make an astronomical discovery, and I say that not just to entice you to read on. Anybody who takes the time to learn well the constellations can be the first to pick out a star that wasn't there the night before. And just a few minutes spent scanning the sky low in the west after sunset, or in the east before sunrise, preferably with binoculars, will, I positively guarantee you, reward the patient searcher with a comet that will forever carry his or her name.

Almost any Nikon, Kodak, or similar home camera can be turned into a powerful discovery machine that will put you right up there among the ranks of well-known discoverers like Bradfield and Evans of Australia, Honda, Seki, and Wakuda in Japan, Austin from New Zealand, Alcock of Great Britain, Meier in Canada, and Levy, Rudenko, and Machholz in the USA.

I have written the following pages in the form of a *Guide*, hoping that it will encourage you and others to partake of the hunt. Since the intended readers obviously have at least a speaking acquaintance with astronomical terminology, I will not use up space defining terms like *magnitude, precession,* or *orbital eccentricity* (although the Appendix will). Similarly, equations and formulae that exasperate some but that others may find interesting have been included – but are confined to the back pages. And finally, while the emphasis of this *Guide* will be on amateurs and the contributions that they can make, we will, nevertheless, look carefully at how the professional discoverers go about their work and learn about the kind of equipment they use.

About Chapter 5, 'The discoverers: in their own words'. If you have time only to read one chapter, read this one. However, I should warn you: it is as long as all the other chapters put together. But if you want advice straight from the horse's mouth, here is the stable.

A disclaimer is needed. A number of commercial products will be mentioned by name, not only by me but by the contributors to Chapter 5. None of us, to my knowledge, have, nor will have, received any benefit thereby from the quoted companies. We name these products simply because these are the products we use. There may well be others better.

Right here and now, I want to take the opportunity to state that I am not a know-it-all. Consequently, I wish to record my very grateful thanks to all the kind, patient, and understanding people who have helped me in this undertaking. First of all there are those that took their time to write contributions to Chapter 5: George Alcock, Rod Austin, Ken Beckmann, Bill Bradfield, Mike and Peter Collins, Bob Evans, Gordon Garradd, Eleanor Helin, Minoru Honda, Albert Jones, Dan Kaiser, David Levy, Don Machholz, Brian Manning, Rob McNaught, Michael Rudenko, Tsutomu Seki, Carolyn Shoemaker, and Tetsuo Yanaka. Translations of the Japanese articles were patiently made by good friends Toru Hayashi and Akira Kawazoe. Additional remarks and information have been incorporated elsewhere in the book; my enlighteners include Jun Jugaku, Carl Pennypacker, and Guy Hurst. Chapter 9 was kindly read critically by Dan Green of the Central Bureau for Astronomical Telegrams. My dear friend, and an amateur in the purest sense, Dr Robert E. Gurley read the entire manuscript critically and made numerous useful comments. It was another good friend, Ben Mayer, who, in the mid-1970s, got me firmly into the discovery game with the PROBLICOM programme, his general unflagging enthusiasm and, most of all, the warm and generous friendship of both Ben and his wife Lou. Gonzalo Alcaino, *mi amigo* and colleague here in Chile, and Director of the Instituto Isaac Newton, has also encouraged my efforts, for which I am very grateful.

Lastly, but certainly not least, my wife Matty stood by me in spirit if not in body during those nights when I was trying my hand at discovery – or ranting at the clouds and streetlights that hampered my efforts.

Note added at proofs:

In the fast-moving world of modern-day astronomy, much has happened since the manuscript of this book was finished. Chapter 5 contributors David Levy and Rob McNaught have increased their discovery outputs immensely by getting access to large Schmidt telescopes. The CCD-equipped Spacewatch telescope of the University of Arizona (see Chapter 8) is suddenly discovering numerous fascinating solar system objects including 1991 VG, which circles the Sun in an orbit remarkably similar to that of the Earth, and 1992 AD with a period of 93 years. And on the purely amateur front, the amazing Bill Bradfield has discovered his 15th comet, 1992b, and in the USA Jean Mueller and Howard Brewington have established themselves as new talents in the field of comet discovery. Another recent entry – from Australia – is Paul Camilleri who is finding novae with breathtaking frequency.

My own efforts have recently included the use of a CCD, and with an SBIG ST-4 (see Appendix J) and a 20-cm Schmidt I have started patrolling southern galaxies for supernovae. No sooner had I begun, when I discovered SN1992A – but did so photographically. Because the supernova was well imbedded in a bright, lenticular-shaped galaxy, it's not clear if I would have picked it up with my new CCD. But the CCD is a remarkable detector and I recommend that sky searchers seriously consider it.

For the record, I should state that this *Guide* was conceived after several good conversations with my friend of many years Simon Mitton of Cambridge University Press. Variable star expert John Isles, now living in Cyprus, was the first person to read any of the original manuscript (the first four chapters) and I thank him for suffering through it and making a number of corrections and valuable suggestions. Another good friend and retired Kodak wizard, Ben Johnson, now living in Puerto Rico, was the last person to assist me. He read the chapter on photographic discovery (7); it became much better as a result.

1

To begin with: initial thoughts and comments

There is no way that professional astronomers can patrol the entire sky nightly. There are just too many square degrees of it (41 253) and too few astronomers – professionals, at any rate – who possess that burning desire to search for new objects. And yet, when something new and intriguing does get discovered, like the 1987 supernova in the Large Magellanic Cloud shown in Figure 1.1, the diminutive asteroid (1991 BA) that came within an astronomical whisker of ploughing into the Earth (Figure 1.2), or Comet Levy (Figure 1.3) that arrived from deep, deep space in early 1990, then the professional machinery shifts into high gear, and the big telescopes with all manner of super-sophisticated equipment are turned towards these newcomers.

With the Hubble Observatory now orbiting the Earth, and the Keck monster telescope in operation, hundreds of professional astronomers are chomping at their respective bits waiting anxiously for something new and entirely different to observe. Whom do they rely on? The humble amateur of course.

Recently my friend and (professional) colleague Gonzalo Alcaino and I were inspecting the CCD image of the globular cluster Omega Centauri shown in Figure 1.4. On this single deep exposure taken with the 2.2-m telescope at the European Southern Observatory, there must have been more than ten thousand stars visible. Very impressive, especially considering the miniscule size of the field, 3 by 4 minutes of arc. Some time later I began to wonder how many similar exposures would be necessary to cover the entire sky. My desk calculator gave me the answer: a little over twelve million. And then I tried what someone once called a 'whatif': What if one hundred professional astronomers observing with one hundred different teles- copes were to expose one hundred CCD images on the same night? What percentage of the sky would be recorded? Answer: Less than one-tenth of one per cent.

To be sure, there is a handful of professionals who are carrying on search programmes, usually with large Schmidt telescopes or reflectors equipped with the latest electronic detectors, but they invariably look up as high in the sky as possible – along the celestial meridian – and they only work a few nights of each month. On a good observing run of four or five nights, they might, with the right equipment, cover a quarter of their available sky. Or they may get completely clouded out.

Fig. 1.1. The naked-eye 1987 supernova (arrow), the Large Magellanic Cloud, and the naked-eye Comet Wilson (circle).

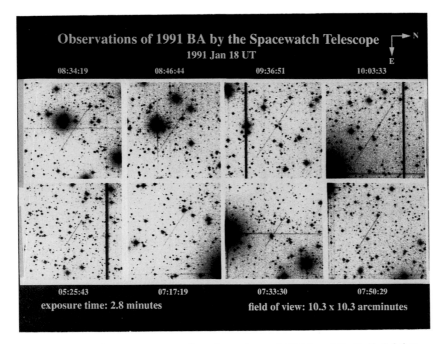

Observations of 1991 BA by the Spacewatch Telescope
1991 Jan 18 UT

08:34:19   08:46:44   09:36:51   10:03:33

05:25:43   07:17:19   07:33:30   07:50:29

exposure time: 2.8 minutes   field of view: 10.3 x 10.3 arcminutes

Fig. 1.2. Asteroid 1991 BA came within a hair-raising 170 000 km of the Earth. It left its trail on this ten-minute exposure made with the University of Arizona's 91-cm Spacewatch telescope on Kitt Peak.

Forty-one thousand, two hundred and fifty-three square degrees. A lot of sky. But not really. Not for your store-bought 35-mm camera with a standard 50-mm lens and a 1000 square degree field (see Figure 1.5). In principle, a person standing on the equator could, in one clear night, photograph all the accessible sky – the half of it not lit up by the Sun, or below the horizon, that is – with 21 half-hour exposures. With the right techniques, stars down to magnitude 11 or 12 would be recorded.

And consider a visual observer manning a telescope – a 6-inch, let's say, with a field of view of two square degrees. He or she could do the same thing taking four seconds to inspect each field. A pair of giant binoculars with a 4-degree field (12.6 square degrees) would permit our intrepid observer to linger for nearly a half minute before moving on to the next field.

Maybe no one person could keep at it all night long, but then there are areas of the sky more endowed with comets, or asteroids, or novae, or supernovae than others. Where are they? The next three chapters will tell.

Fig. 1.3. Comet Levy (1990c) lived up to its expectations developing a fine tail faintly visible with the naked eye. Photo by Gordon Garradd with a 25-cm f/4.1 reflector using Kodak's Technical Pan film.

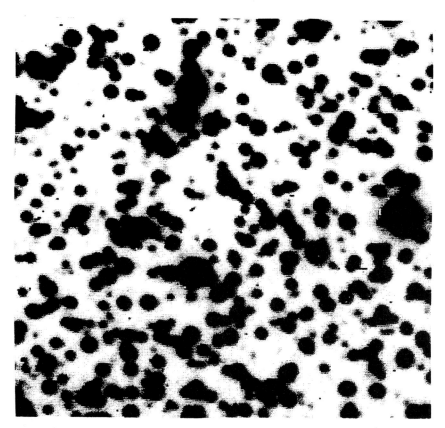

Fig. 1.4. A small section of Omega Centauri, the brightest globular cluster visible from Earth, as imaged by a large telescope – the 2.2-m Max-Planck-Institut reflector at the European Southern Observatory, and a modern CCD. Taken by Gonzalo Alcaino and the author.

The prey

What should be searched for? Comets, naturally. Novae and supernovae, certainly, but supernovae in *other* galaxies, not ours. Discovering asteroids generally requires a photographic telescope, but the ones that come close to us, the Earth-threateners, can get bright enough to be seen with binoculars. And finally variable stars. Many brighter than 9th magnitude are still waiting to be found.

These five types of objects are our primary prey, and they will occupy most of our attention in the following chapters. But first, a few words about each.

Someone once estimated that on average, at any given moment, there are six comets up in the sky brighter than 12th magnitude, and an estimated 60 per cent – or 3.6 of them – have not yet been, and possibly might not ever get discovered. They

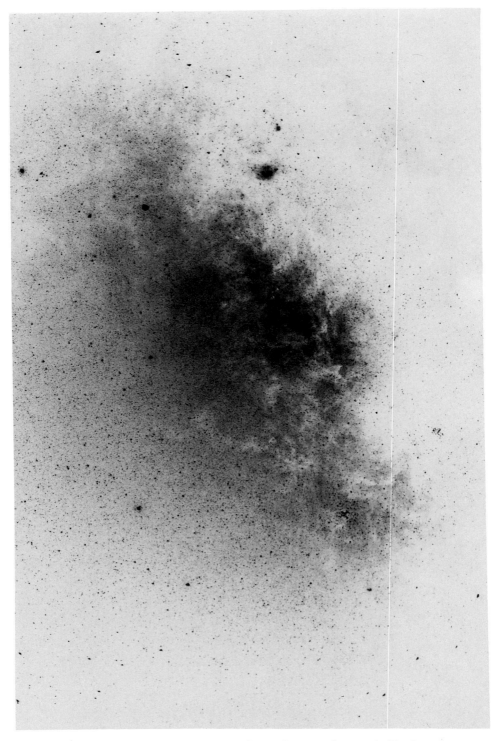

Fig. 1.5. Sagittarius and the brightest part of the Milky Way, photographed by the author with a 35-mm camera.

are just sitting there, like proverbial ducks on a pond, waiting for someone to come along and spot them. As comet discoverer Michael Rudenko reasoned to himself so clearly, 'If I could see newly discovered comets, it should be possible for me to discover them.' If you have a telescope or a camera that can reach to 12th magnitude, you should ask yourself: Why not start looking?

As for novae, Ben Mayer and I once wrote an article in which we estimated that every year an average of 15.9 novae burst forth and reach magnitudes brighter than 11. To be sure, some of them will never be seen because at times the Sun and the Moon get in the way, which brings the total down to about 11 discoverable novae per year. But the considerate behaviour of novae, more than any other of our primary prey, is that they are nearly all found in one narrow sliver of sky, the Milky Way.

Supernovae. To make a gross understatement, these rare events are best looked for in other galaxies. But because of their immense intrinsic luminosity, they can be seen at great distances from the Earth. The 1987 supernova in the Large Magellanic Cloud 160 000 light years away, reached magnitude 2.8; a supernova in the Andromeda Galaxy, more than ten times further away, could (and did) become visible to the naked eye; and in the populous cluster of galaxies in Virgo, supernovae regularly peak at 12th or 13th magnitude – and usually several blow up every year.

And the asteroids. Most of these minor planets keep to themselves, moving in orbits between those of Mars and Jupiter, and most of the brighter ones have now been discovered. But the fainter one looks, the more asteroids there are. While professionals still make most of the discoveries of new 'main belt' asteroids, amateurs continue to add new members to the rapidly growing roster of asteroids. More exciting, of course, are the Earth-threateners, and every year one or two can get bright enough to be seen with modest equipment.

Myriads of undiscovered variable stars of many different types, many within range of 35-mm cameras, are known to exist, and here amateurs are leading in the department of discovery. The British variable star specialist, John Isles, estimates that the current discovery rate is almost one a night. Hundreds a year!

There are other objects, of course – unexpected meteor showers that have, on rare occasion, been described 'like falling snow' (Figure 1.6), meteorites that plunge to the Earth without warning, unpredictable displays of the Northern or Southern Lights, and changing features on the visible surfaces of the Sun and major planets. In later chapters we will have something to say about them all.

The discoverers

Here we encounter some interesting curiosities. In 1987 Brian Marsden, Director of both the Central Bureau for Astronomical Telegrams and the Minor Planet Center, pointed out that since 1919, 'Essentially all the amateur contributions (discoveries) ... have been made by fewer than a hundred individuals, not one of whom is a female.' The last phrase is especially puzzling: two of the currently most successful

Fig. 1.6. The famous wood-cut of the Great Meteor Shower of 1733.

Fig. 1.7. 'Tsutomu Seki of Kochi, Japan, emerges quite definitely as the leading all-round amateur astronomer in the world' (Marsden, 1987). Here he is.

professional comet hunters are Carolyn Shoemaker and Eleanor Helin of the USA; and one of the more prolific supernova disoverers is Natal'ja Metlova in Russia.

Marsden also notes that almost all the recent non-professional discoveries have been made in a handful of countries. To update and paraphrase Marsden, 'Between 1919 and 1991, there have been 19 amateur astronomers who have discovered (or co-discovered) three or more comets. Curiously, these individuals represent only eight different countries: Japan, USA, Australia, Republic of South Africa, UK, Canada, New Zealand, and Lithuania.'

The amateur activity in Japan continues to be impressive. In discoveries of comets, asteroids, novae, and supernovae, Japanese amateurs either lead the list or come second, with two names, Honda and Seki, pre-eminent in three of the categories. For many years, the late Minoru Honda led in discoveries of comets and novae; Seki, who currently devotes most of his time to hunting asteroids and measuring their positions, was called by Marsden (in 1987) the leading all-round amateur in the world (see Figure 1.7). The weather of Japan, it should be noted, does not generally provide exquisite observing conditions.

The secret ingredient

South Australian (ex-New Zealander) William Bradfield told me that, in his estimation, the weather is the biggest problem in comet hunting. From the vicinity of oceanside Adelaide, he has found, so far, a record (for 20th century amateurs) 14 comets; imagine what he might have done had he lived in southern Arizona or northern Chile! Clear skies certainly are a boon to comet searchers, but some, like George Alcock in Central England (five comets), Rolf Meier of Ottawa, Canada (four), and Michael Rudenko of Massachusetts (three), to say nothing of the numerous successful comet, nova, and asteroid hunters in Japan, must suffer with skies that are far from perfect. The great Leslie Peltier (ten) lived in central Ohio where the weather is so unreliable that the state's two most important telescopes were eventually moved away to sunnier (and starrier) Arizona. And Reverend Robert Evans, who has found, up to now (October 1991), an almost unbelievable 22 supernovae, lives on the East Coast of Australia where the skies frequently leave sky searchers with much time to read.

So, what is it they all have in common to make themselves so successful at the hunt? The answer can be found in the lines – and between the lines – that these successful discoverers have written (see Chapter 5). They all have *persistence*. While most are ready with anecdotes about the comets or novae or supernovae that they should have discovered and missed, and while some comment freely on the number of hours of searching that it took to find a comet, they rarely, if ever, give up. *Stick-to-itiveness* was the word that my good friend and sometime discoverer, Art Hoag used. Fruitless nights that might put an end to the hopes of others do little to deter the efforts of those who have been most successful.

Benefits for the discoverer

Why do they do it? Probably for nearly as many reasons as there are discoverers. Money is certainly not an incentive. William Herschel received a life pension and knighthood after he accidentally found what he thought might be a comet but in reality was Uranus. But then Uranus could be considered to be the first major planet *ever* to be discovered – and that was back in 1781. Also, Herschel had the good sense to propose that the new planet be named 'Georgium Sidus' after the King of England. If you were to discover the tenth planet, do you think that trying to name it 'Elizabeth', or 'Bush', or 'Akihito' would get you anything more than a headline in the newspaper and a nice thank-you note?

Nevertheless, a modicum of fame does accompany the discovery of something truly awesome like a brilliant comet or a nearby supernova. After Luboš Kohoutek found what was touted to become 'the comet of the century', he was invited on a luxury Caribbean cruise ship to give lectures and show the wondrous new sight to the other passengers. But several months after Rodney Austin discovered the most recent 'comet of the century', he wrote to me that virtually nothing had changed in his life. Even in his own country there was little reaction or rejoicing.

On the other hand, my Japanese friends tell me that Minoru Honda achieved virtually a cult status after his numerous comet and nova discoveries. He, perhaps more than anyone else in Japan, is responsible for the enthusiasm for sky searching that continues in that country. And one should not forget that having one's name appended to a comet or an asteroid does, after all, win one a virtual immortality.

More important to today's discoverers is just the plain satisfaction of finding something new. In what other field can a neophyte make a 'scientific discovery'? Perhaps botany or entomology, but that would almost certainly require travelling to and living in some remote jungle area.

There is a delicious rush of pleasure in knowing that you are probably the first person in the entire world – *perhaps in the entire universe* – to see a newly found celestial object. When the realisation first comes that that star-like object or fuzzy patch is something new, the pulse quickens and the adrenalin begins to flow freely. Staid professional astronomers tend to disparage this type of 'backyard' activity, but speaking as an ex-professional – a born-again amateur if you will – I can attest that professionals owe much to the amateurs who find new and intriguing things for them to devise obscure theories about.

Endpiece

If you were to ask me, at this early juncture, for the one most valuable piece of advice I can give (besides exercising persistence), I would say *Know Your Prey*. Read up on it in a good up-to-date textbook, and keep abreast of new developments and recent discoveries through magazines, the IAU *Circulars*, newsletters, or one or more of the computer bulletin boards. Successful discoverers are not, as a general rule, active in clubs and societies; they just don't want to lose valuable dark time socialising at star parties and the like. Nevertheless, many organised groups have newsletters and journals, and through them you can meet others interested in discovery or in your own search project. And it's a good idea to have loyal, well-equipped friends to confirm your discoveries.

2

Comets and their kin: general characteristics

There are, of course, two kinds of celestial hunters in this world: those who leave it all to chance, and the successful ones, those who know at least a little something about the general behaviour of their astronomical prey, and have a good idea as to where to look, when to look, how, and with what. This chapter, mainly about comets, and the next two are for those wishing to join the second group.

We will be concentrating on those five types of celestial objects most searched for: comets, novae and supernovae, variable stars, and asteroids. Other interesting phenomena such as the Sun and planets, meteor showers and meteorites, and the aurora will not be forgotten; they will receive somewhat less attention. In this and the next two chapters the aim will be to review briefly the behavioural characteristics of these celestial bodies and phenomena, and throughout all the chapters, we will continually intersperse sage words and useful commentary regarding search strategies for each class of object.

No longer are comets and novae discovered visually from Paris rooftops *à la* Messier and Pons. Observers and their search techniques continually change with time. To keep us as up-to-date as possible, we will thoroughly analyse and comment on the veritable flood of astronomical discoveries made during the decade of the 1980s. In this way, we can learn from the experiences of currently successful hunters, identify under-patrolled regions of the sky, and note types of objects best left for the professional to go find.

Comet nomenclature

Early in the morning of New Year's Day, 1989, Japanese amateur Tetsuo Yanaka, observing with 25 × 150 binoculars, picked out a slowly moving 11th magnitude smudge some 29° above his horizon and about 11° southwest of Arcturus. As he later was to learn, it was the first comet discovery of the year and initially became known as Comet Yanaka (1989a).

Sixteen and a half hours later in one of the smaller domes on Mount Palomar, Eleanor Helin and some colleagues began an exposure with a 46-cm Schmidt camera. Later inspection of the film revealed the year's second comet, Comet Helin–Roman–

Fig. 2.1. Comet Austin ($1989c_1$). Even though it did not live up to early expectations, it was the brightest since Comet West 1976 VI. Photographed with the 46-cm Palomar Schmidt by the Helin Team at JPL. (See Chapter 5.)

Crockett (1989b). (Other members of her team might have been involved, but they had to be left out since a comet's name can list no more than three discoverers.)

During the year 1989 a total of 29 different comets were seen for the first time, some were new discoveries, and others recoveries of periodic comets with well-known orbits. After the zth comet was found, the lettering started anew using now a_1, b_1, etc.; Comet Austin ($1989c_1$), shown in Figure 2.1, was the last (and the brightest) of the year's harvest.

The year's fourth comet was Periodic Comet Russell 3 (1989d), or simply P/Russell 3; it became the first previously known comet be to recovered. The number 3 indicates that it was the third 'short-period' comet (see next section) discovered by K. P. Russell at Siding Springs in Australia; he first identified it in June of 1982.

Later, after precise orbits could be calculated, it was determined that Comet Yanaka (1989a) was the 20th comet to have passed perihelion in 1988, and it now and forever carries the name Comet Yanaka 1988 XX, although the first name will, on occasion, also be used. P/Russell 3 continues to be known by its earlier name, Comet 1982 IV. Final word is not yet in on 1989b.

Orbits

My most recent edition of Brian Marsden's valuable *Catalogue of Cometary Orbits* – the seventh – contains orbital parameters for 1292 cometary apparitions observed through the year 1990. As Marsden notes in the introduction, these appearances were made by 810 different comets. Of these, 155 – 19 per cent – are classified as 'periodic', that is, having periods of less than 200 years. A look at the majority, the 81 per cent that are non-periodic or long-period comets, reveals that their orbits have eccentricities, *e*, very close to 1.0, meaning that their orbits are extremely close to being parabolic. (See Appendix E for further remarks on orbits and their parameters.) This tells us that these comets have come into the neighbourhood of the Earth–Sun system from immense distances away and will, for all practical purposes, leave our planetary system forever; their periods are essentially infinite. In fact, some are. However, most will probably settle back into the distant Oort Cloud of comets located thousands of astronomical units away from the Sun, but that is a topic we will not pursue here.

Of more interest is the realisation that these long-period comets have never been seen before, at least not in historical times. They had to be discovered.

In reality, the only orbital characteristic that all the long-period comets share is an eccentricity close to unity. Consider one easily visualised quantity, the orbital plane inclination, *i*. If we tabulate this angle for all the long-period comets listed in Marsden's *Catalogue*, we find inclinations of all values but with a mild tendency for high-inclination orbits. (A bar graph of these inclinations appears in Figure F5 of Appendix F. The curious peak at 145° will be discussed in a moment.) Probably, some of the low-inclination orbits have, over the eons, been 'cleaned out' by the gravitational attractions of the major planets.

Conclusion: the search for long-period comets can, in principle, be made anywhere in the sky. However, see Appendix F, and especially Figure F1, for the details of the discoveries of the 1980s.

To this advice should be added the not fully understood, but long recognised fact that a little over 60 per cent of all long-period comets get discovered in the morning sky. According to comet statistician Edgar Everhart, three-quarters *could have been* first observed in the morning sky. He suggests that the reason for this anomaly has to do with the direction of the Earth's motion in its orbit: on the front side of the Earth (the morning side) we are closing in on the comets more rapidly and, consequently, they brighten up more quickly. However, others aren't so sure. It may be that comet hunters, refreshed after a night's sleep, and finding observing conditions more tranquil, work more efficiently. Whatever the real reason, one thing is clear: you will discover more comets in the morning than in the evening.

As for the short-period comets, they presumably got that way because sometime in the past they were influenced by the gravitational tugs of the larger planets. Most of these 155 comets have periods around six years (see the graph of Figure F6 in Appendix F), and nearly all lie between five and ten years. Close encounters with the major planets, mainly Jupiter, occur upon occasion; consequently, the roster of short-period comets is always changing, albeit slowly. Still, they get discovered in moderate numbers. For example, 11 were first sighted in just the two years 1987 and 1988, and four were 13th magnitude or brighter, which is to say, discoverable by modest means. Comet Ciffreo 1985 XVI was 10th magnitude at time of discovery; its period is 7.2 years.

Another noteworthy characteristic of the short-period comets is that all but five move in direct orbits, i.e., in the same direction as the Earth and the other major planets. (One of these five is Halley's Comet.) No comet with a period less than ten years has retrograde motion. And there exists a strong preference for low-inclination orbits; almost half have orbital inclinations of less than 10°. (See the graph of Figure F5 in Appendix F.)

The appeal of discovering a reasonably permanent short-period comet that will bring 'your' discovery back to the solar neighbourhood at frequent intervals can be satisfied, then, by concentrating the search in regions close to the ecliptic and again in the morning hours where the Earth catches up with these more slowly moving short-period comets.

Behaviour

At a distance from the Sun of more than a few astronomical units (AU), a comet is effectively in deep freeze. Not much gas or dust will escape to surround the dirty, icy core. It looks and behaves much like an asteroid or any other solid body (see next section). But by the time a comet arrives within a couple of AU of the Sun, the warming action starts to melt the ices, and the comet begins to change its form: around the nucleus a cloud of gas and fine dust appears – the coma – and if the comet is sufficiently large, a tail begins to sprout. The tail continues to grow, sometimes not so steadily, as the comet approaches perihelion, reaching its maximum length some days after closest approach. Thereafter, it shrinks away, eventually returning to the celestial freezer only slightly the worse for its warming experience – unless it passed very close to the Sun and was subjected to the intense tidal forces of the Sun.

In reality, every comet has two tails (some say three), one composed of ordinary gases like carbon dioxide and water vapour, the other of microscopic solid particles – sandy stuff and other minerals. As Figure 2.2 shows, the two tails can appear separated in the sky, blown back from the comet's head by separate forces, the gas by the solar wind and the dust by the pressure of sunlight. Also, the colours differ: the carbon dioxide, glowing from the action of ultraviolet sunlight, shines with a bluish tinge, while the dust normally has a whitish, yellowish, or even reddish hue. These subtleties, usually not noticeable visually, show up nicely on colour photographs as Figure 2.2 illustrates.

Until a sufficient number of accurate positions have been measured to allow an accurate orbit to be determined, there is little to distinguish a new short-period comet from one with a long period, at least not at first. But with every approach to

Fig. 2.2. When the orientation of the comet is just right, the gas and the dust tails of a bright comet can be well separated. Halley's Comet, March 17, 1986. Photographed by the author from Easter Island as a part of NASA's International Halley Watch.

the Sun, a comet suffers a certain amount of meltdown and consequent mass loss from the warm sunlight. The result is a cleaning action of these 'dirty icebergs', the effect of which is to take away some of the dusty material. This loss, if great enough, can be noticed both in the coma surrounding the nucleus and, of course, in the tail. Don Machholz had the good or bad luck, depending on how you look at it, to discover a periodic comet (1985e) that seems to be on its last legs. P/Machholz 1986 VIII comes closer to the Sun (0.13 AU) than any other periodic comet, and does it frequently, once every 5.25 years. Both professionals and amateurs will be watching it closely at every perihelion passage looking for further signs of decay.

A dozen or so comets were thought, at time of discovery, to be asteroids because so little gas and loose dust remained. And as we will see in Chapter 4, some asteroids are believed to be totally 'de-gassed' comets. For sure, a certain kinship develops between the two types of objects when comets grow old and lose their splendour.

Brightness

A solid body, like an asteroid (see Chapter 4), brightens up inversely both as the square of the distance from the Sun and the square of the distance from the Earth: halve either distance and the brightness increases by a factor of four. A comet does better, however, because as it warms up, gas and dust are released to form the coma and the tail. As a result, a young, long-period comet will brighten up inversely as the cube or even fourth power of the distance to the Sun; halve the distance, and the comet brightens eight or sixteen times. Quite a difference. And if there is much frozen volatile material and the comet is not well bonded together, it may behave according to the fifth or sixth power: a 32 or 64 times increase in brightness every time the solar distance is cut in half. Halley's Comet brightens up as the fifth power of the solar distance.

These relationships can be put into convenient-to-use formulae; in Appendix C, you will find them together with sample calculations. After a new comet is discovered, the guessing game begins: which law of brightening will it follow? Usually, a conservative fourth power is chosen to predict future brightness, but the comet may not even live up to this level of expectation. Comets like Kohoutek in 1973 and Austin in 1990 might have been spectacular objects had their icy inner parts not been so crusty. As it was, they flared up briefly when they first felt the warming action of the Sun and the few volatile gases at the surface of the nucleus were burnt off. But then these two comets continued their trip towards the inner solar system following a brightening law close to a disappointing fourth power.

Still, brightening up as the fourth power signifies a considerable increase in visibility, illustrating clearly why the best place to search for a comet is as close to the Sun as is practicable, especially if your telescope or camera is modest in size. Consider the 37 'amateur comets' discovered during the decade of the 1980s: all but eight were found within 90° of the Sun.

Many readers will remember Comet Ikeya–Seki in 1965; it came perilously close

17

Fig. 2.3. Comet Ikeya–Seki 1965 VIII, the magnificent 'search light' comet, and the best 'Kreutzian' comet of the century . . . so far. Photo by the author with a 50-mm lens and Tri-X Pan film.

to the Sun and put on a spectacular display. Its predicted and estimated magnitude at perihelion was an incredible − 10, visible in broad daylight. Shortly after perihelion, the tail stretched more than 60° across the sky. (Figure 2.3.)

Earlier we mentioned that there was an excess of comets with an inclination averaging close to 145°. Comet Ikeya–Seki is one of these 'Kreutzian comets' (named after the man who first noticed this grouping); its $i = 141.9°$. At closest approach to the Sun, Ikeya–Seki was just over a million kilometres from the Sun's centre, or about 1.7 solar radii. According to a suggestion made by Brian Marsden, once upon a time there was a monster comet in a very similar orbit, but the close passage of the comet by the Sun was too much for it, the immense tidal forces tore it into shreds. One of the remnants was Comet Ikeya–Seki – and over the years others have been seen. In the 1980s, 15 Kreutzian comets were discovered but, sad to say, none were seen or photographed from down here on Earth: they were all detected by orbiting coronagraphs that were able to pick up some of the smaller members of this group as they passed through the outer solar corona.

But another Kreutzian comet the size of Ikeya–Seki – or larger – is probably

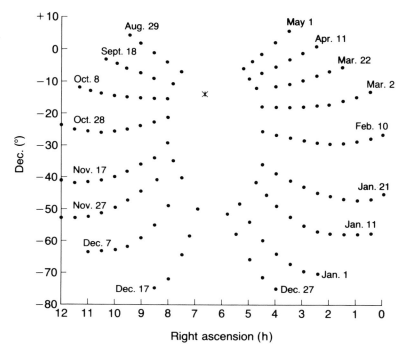

Fig. 2.4. The family of arcs defined by the common orbit of the 'Kreutzian' comets. On a given night, in any year, there exists one arc anywhere along which there could be a potentially spectacular comet such as Ikeya–Seki.

heading in towards the Sun even as you read this. Knowing its orbital elements in advance gives us some excellent clues as to where and when to look. Consider the following: at any given moment, an orbit can be plotted on a map of the sky; that is to say, if the orbit were filled with luminous material, it would mark out a sweeping arc through the stars. Because the location of the Earth in its orbit is continually changing, the position of this arc never stays put, but we can, nevertheless, plot the location of the orbit showing where it would lie relative to the stars at, say, weekly intervals.

The family of arcs for Kreutzian comets looks like Figure 2.4. Find the date on which you will be observing, and you have defined a specific pathway through the sky somewhere along which a potentially spectacular comet may be moving.

One last comment concerning short-period comets. Earlier, we noted that these comets tend to stay close to the plane of the ecliptic, and that they usually can be expected to brighten up more rapidly than long-period comets. Thus, our earlier suggestion to look for short-period comets a little east of the anti-Sun direction is sound strategy. (Also, read about 'The discovery gap' below.)

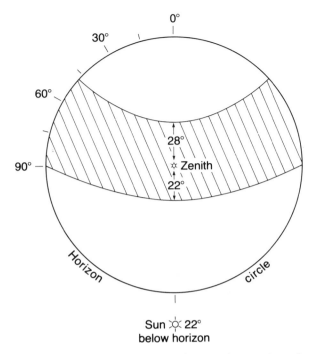

Fig. 2.7. The region in the sky corresponding to 'the gap' of Fig. 2.6 (cross-hatched). The magnitudes of the comets should average around 12 or 13.

The remarkable conclusion that one can draw from Figure 2.6 is that there are not many comets being discovered at distances between 90° and 140° from the Sun. Of the grand total of 59 comets discovered by the world's amateurs and the Palomar pros, only seven – 12 per cent – were found within this gap.

Shouldn't this be a good place to begin searching? It seems unlikely that the gap is caused by a paucity of comets; more likely, it exists simply because no one is looking there: amateurs don't search that high in the sky, and professionals don't search that far from opposition. The crucial question is how faint you can work. The average magnitude of all the amateur comets was 9.9; that of all the Palomars, 14.6. Therefore, one would expect the comets in the gap to fall between these two values. The average magnitude of those seven in the gap is 13.4 with a range from 7 to 17. But before you set your sights accordingly, you should read carefully the contributions by Eleanor Helin and Carolyn Shoemaker, as well as the section in Chapter 7 describing the Palomar search programme.

Exactly where in the sky is this gap to be found? Figure 2.7 shows where at a time about two hours after sunset or before sunrise. What on Earth [sic] could be easier?

3

Novae and supernovae: nuclear runaways

We now move out of the solar system to consider two classes of what some have called 'cataclysmic' variables, stars that suddenly erupt in brightness with little or no warning. The novae and the supernovae, in a class by themselves, are the most spectacular and luminous members of this group. We will devote this chapter to them.

Part 1: Novae
Nomenclature

Although novae can be seen in many galaxies besides our own, most of the ones that we will be interested in have been and will be discovered in our own Galaxy. Their designations include the name of the constellation (in the genitive case) in which they were found and the year of discovery. Hence, when the ace Japanese nova hunter, M. Wakuda discovered a new star in Cygnus in June of 1986, it became known as Nova Cygni 1986, or simply N Cyg 1986. If more than one nova is found in a given constellation in the same year, then Arabic numbers are appended; Rob McNaught found N Cen 1986 No.2.

The International Astronomical Union has assigned the Moscow Bureau of Variable Stars the task of giving official and permanent names to all newly discovered and confirmed variables, including novae. N Sco 1967 is now and forever known as V916 Sco, and N Lib 1983 has officially become GW Librae. The naming system for variables in general is complicated; here we will only say that the genitive case of the constellation name follows double Roman letters (except for R through Z) for the earliest variables discovered. When the letters run out, then designations begin with V335, V336, etc.

Cause

Novae and supernovae are totally different beasts. To begin with, a nova's explosion is puny compared to that of a supernova – in round numbers, ten magnitudes punier. The nova outburst is the ultimate result of a steady dribble of material that falls on the surface of a compact white dwarf star from an exceedingly nearby companion star. After a period of time, a critical amount of this material accumulates on the white dwarf. This star's surface gravity is so incredibly high that nuclear reactions then take place spontaneously, and Bang! Off goes the nova. Full rise in brightness averages 10 or 12 magnitudes, or even more; the size of this increase in brightness varies considerably from nova to nova. (More on this topic later.)

Often the pre-nova system can be seen as a faint star on photographs taken before the outburst. For example, a 9th magnitude nova might have a magnitude of 19 or 20. As we will see in the next section, this presence can provide important information.

At maximum brightness, an energetic nova will have an intrinsic luminosity rivalling the most luminous stars in the sky. Consequently, they can be seen at great distances from the Earth. The novae in our distant neighbour the Andromeda Galaxy (M31) frequently get as bright as 16th magnitude.

Frequency

Dozens of novae go off every year in our own Galaxy, but we live in a large and dusty island universe, and in the spiral arms, among the myriad of stars, there exist great quantities of obscuring matter, dark examples of which can be seen as irregular patches in the Milky Way. (See Figure 3.1.) As a result we only see those novae that occur in our part of the Galaxy; most of those that reach magnitudes of 12 or brighter at peak lie within 10 000 light years of us.

In a bountiful year as many as a half-dozen new galactic novae might get discovered. During the 1980s, 33 novae were discovered with nothing more than ordinary 35-mm cameras or by dedicated visual observers who were able to recognise the presence of a 'new' star. (See Table F2 in Appendix F.) To this number we can now add the two or three novae per year that are being discovered with only slightly more sophisticated cameras in our two closest galactic neighbours, the Magellanic Clouds visible, alas, only from the tropics and further south.

Special mention must be made of the so-called recurrent novae. Because only a very small fraction of one per cent of the available stellar material is involved in a nova outburst, little will have changed after an eruption, and the whole process can begin again. Intervals between explosions vary greatly, but it seems to be a general rule that the more violent the outburst, the greater will be the time before the next explosion. (See Figure F13 in the Appendix). The best-known recurrent nova, T Pyxidis, goes off every 15 or 20 years, rising from an inconspicuous 15th magnitude up to 6th or 7th magnitude. VY Aquarii, called a dwarf nova by some because of its comparatively mild outbursts, was seen to erupt four times in the 1980s. In the next chapter, we will have more to say about VY Aqr and related types of cataclysmic variables.

Behaviour

Thanks to two serendipitous series of photographs taken by amateurs Peter Garnovich in Maryland and Ben Mayer in California, we know that N Cyg

Fig. 3.1. Dark clouds of obscuring material in the Milky Way in Aquila. Photographed by
the author with a 85-mm lens.

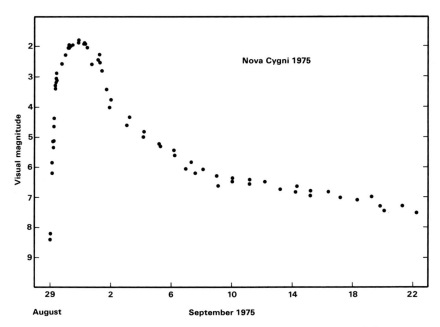

Fig. 3.2. The light curve of Nova Cygni 1975. All the observations were made by amateur astronomers, mostly members of the AAVSO.

1975 = V1500 Cygni rose steadily from magnitude 8.4 to magnitude 2.4 in just under 24 hours. Peaking at magnitude 1.9 two days later, this newcomer dropped quickly back to magnitude 4, and afterwards declined in brightness more slowly – at a rate of about a magnitude a week. The light curve shown in Figure 3.2 is fairly typical, but some novae linger on longer than others. We frequently distinguish novae by their rate of decline: a typical *fast* nova will drop off at a rate of a magnitude a week or even more rapidly, whereas *slow* novae fade daily by only one or two hundredths of a magnitude.

Knowing about rates of decline in brightness is vitally important when nova searching. Obviously, if skies are scanned only once a month, many fast novae will be missed, an unfortunate circumstance, not just because of the missed opportunities, but because the fast novae are the more energetic and therefore the more interesting to professional pundits.

Where in the sky do novae occur most frequently? Answer: Where stars occur in greatest profusion, and that means the Milky Way. But not just anywhere in the Milky Way. Since novae involve white dwarfs, stars that are highly evolved and that have used up their nuclear fuel, we should, therefore, pay scant attention to those areas where star-making is still under way, dusty and gaseous regions like the Orion Nebula or the Trifid in Sagittarius. However, remember that these well-known

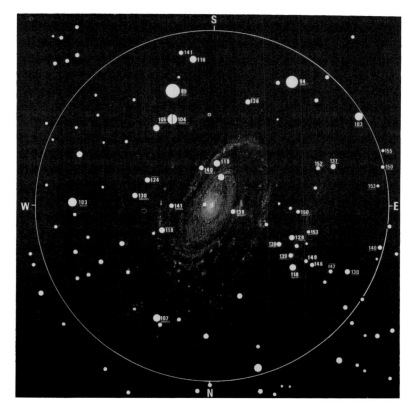

Fig. 3.3. A schematic diagram of our Galaxy as seen from the North (galactic) Pole.
From the Thompson–Bryan *Search Charts* (see Appendix H).

nebulae are relatively nearby – a mere few hundred light years away – and we should
be much more interested in rich star clouds thousands of light years distant. At these
greater distances, we can pack in more stars per photograph, thereby raising our
chances of finding a nova.

Sagittarius holds the constellation record for most novae per year. In this direction
we are looking towards the nucleus of our Galaxy from our location in a spiral arm
about two-thirds of the way to the Galaxy's edge. (See Figure 3.3.) Because we look
across a relatively clear gulf between arms to an inner spiral called the Sagittarius
arm, we get a fairly clear view of the goings on in one of the richest star fields in the
Galaxy. And looking at right angles to this direction, longitudinally down through
our own spiral arm in the directions of Cygnus and Carina, we also see great numbers
of stars. However, in these directions, interstellar matter begins to obscure the view
beyond a distance of a few thousand light years.

It is revealing to study the salient characteristics of novae recently discovered, and prospective nova seekers should devote ample time to studying Table F2 and Figures F10 to F13 in the Appendix. Note that five of the novae of the 1980s reached naked-eye brightness, i.e., magnitude (mag.) 6.5 or brighter. Most novae showed up on patrol photographs taken with 35-mm cameras. At least two dedicated sky-watchers, Peter Collins and Kenneth Beckmann, made their discoveries visually, and in Chapter 5 they tell how they did it.

As you would expect, the fainter one searches, the more one finds. (See Figure F12 in the Appendix.) Some novae fainter than 10th or 11th magnitude probably don't get reported, and they should receive more attention; they could be novae on the rise or, a more important possibility, they could be distant supernovae (see next section).

Where to search? Answer: The entire Milky Way but especially from northern Cygnus through Aquila, Sagittarius, and Scorpius, on past Norma, Crux, and Centaurus to Carina. Figures F10 and F11 show where the novae of the 1980s were found.

Observers in the southern hemisphere are definitely not keeping up with their northern colleagues. Sometimes, several weeks will go by without any search being made of the gorgeously rich southern Milky Way. Since the only person I know of that has a regular programme of nova search in the south is me (as of this writing), I can attest to this fact. My wife and I occasionally do take vacations. Someone, please help out down here.

Note that most of the novae of the 1980s occurred within 15° of the galactic equator; all but five brighter than 11th magnitude fell within these bounds. Obviously, your discovery rate will fall off rapidly if you venture beyond this latitude. But, on the other hand, because interstellar dust is less bothersome at such high galactic latitudes, and because the stars there are generally closer to the Earth, the novae will usually be brighter than their low-latitude kin.

And remember that the Milky Way regions with rich star clouds will be the best hunting grounds. You should know that 16 of the 33 novae 10th magnitude or brighter – almost half – occurred in a 30° by 30° box centred on the galactic nucleus in the same direction as, but far beyond, the 4th magnitude star 3 Sagittarii. Photographers should keep this star in mind as a possible guide star.

For photographers, there is one very serious problem with searching through fields crammed with stars: their sheer quantity. (See Figure 1.5.) In Chapter 7 we will describe a clever way to cut down on the immense numbers that show up on deep photographs of the more densely populated regions of the sky.

Part 2: Supernovae
Nomenclature

The naming system for supernovae – SNe – is rather like comets, using again letters, this time capitalised, after the year: thus SN 1987L was the 12th supernova discovered in 1987 (and the first for Californian Dana Patchick). But because since 1604 supernovae have been found *only* in other galaxies, the name of the galaxy is frequently included for clarity and convenience, and so we would more likely read 'SN 1987L in NGC 2336'. After the 26th supernova of a year, the lettering begins again with aa, ab, ac, etc.

Historical supernovae in our Galaxy carry special names, like Kepler's, or Tycho's, and the nebulae or radio sources that remained after the explosions have their own special names, like the Crab Nebula (alias Messier 1, NGC 1952, Taurus A, and Taurus X-1), or the radio source Cassiopeia A (also faintly visible on deep photographs).

Cause

The supernova explosion, catastrophic, or nearly so, to the star, results from the collapse of a very massive star that follows when it runs out of nuclear fuel. It's rather analogous to the collapse of a human pyramid when one of the poor guys at the bottom gives out: when there is no more energy being generated at the star's centre, the outer layers fall in. Some supernovae (Type II) appear to be isolated supergiant stars that evolve naturally to this brilliant flash finish; others (Type I) are theorised to follow the pattern of novae involving a white dwarf, but ones that grow disastrously from an immense over-accumulation of material from a close-by companion star. Whereas novae recur over and over again, a supernova is reduced once and for all to a super-compact neutron star and a rapidly expanding cloud of gas that once made up the star. Sometimes no visible star is left at all.

Supernovae, roughly 10 000 times (ten magnitudes) more energetic than novae, when at their brightest, can outshine an entire galaxy. See Figure 1.1, for instance. In terms of search strategies, this strongly suggests that if you can see or photograph a galaxy, you surely will be able to see or photograph most of its supernovae.

Frequency

For reasons only partially understood, supernovae seem to occur more often in some galaxies than others; estimates of frequencies in a galaxy the size of ours range from one every ten years to one every two or three centuries. The record holder as of now is the barred spiral M83 (NGC 5236), shown in Figure 3.4. This handsome 10th magnitude galaxy just north of the Hydra–Centaurus border has produced at least six supernovae since 1923, the most recent in 1983. Three other galaxies, M58 (NGC 4579) in the Virgo Cluster of galaxies, NGC 1316 in Fornax, both tightly wound spirals, and the bright barred spiral NGC 1559 in the southern constellation of Reticulum, each produced two supernovae during the 1980s.

In Table F3 of the Appendix you will find listed all supernovae discovered in the 1980s that reached magnitude 15.0 or brighter and the galaxies in which they

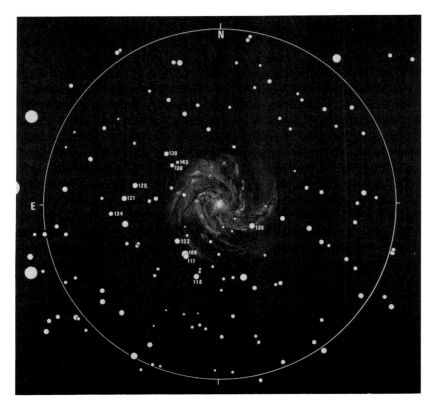

Fig. 3.4. M83, an SBc galaxy in Hydra, produces a supernova at least once every 12 years. There must be dozens of galaxies just like it. From the Thompson–Bryan *Search Charts*.

occurred. At first glance it would seem that spiral galaxies (S types) are the favoured location for outbursts, but this may only be an illusion; among the brighter galaxies in the sky, spirals are by far the most common. (See Figure F16.) Also, for various reasons they probably receive more attention from supernova hunters. But some experts believe that in actual fact, the armless ellipticals (E types) are just as prolific, but then much of an elliptical galaxy is so crammed with stars, it would be difficult to pick out a newcomer even if very luminous. My advice: if you just want to find supernovae, concentrate on spirals, and of course, the bigger and the closer, the better. Leave the ellipticals to the professionals. If, on the other hand, you want to be a truly unbiased investigator, include ellipticals – but don't count on as many discoveries.

During the 1980s, one person, the indefatigable Reverend Robert Evans of Australia, discovered an astonishing 18 supernovae all visually and most with nothing more than a homemade 10-inch reflector. And he is still going strong. There is more from (and about) Rev. Bob in Chapters 5 and 6.

The last *galactic* supernova to have been seen was first noted by the great astronomer–mathematician (and professional astrologer) Johannes Kepler in 1604, although the supernova that produced Cassiopeia A was almost certainly visible to the naked eye (but for some reason not seen) in the late 17th century. One might say that we are overdue for another galactic supernova. However, despite the immense luminosities of supernovae, the presence of interstellar matter drastically cuts down our range. We are probably unable to pick up more than a tenth of all that occur in our home spiral system.

Behaviour

The most recent *naked-eye* supernova – and the last one since Kepler's (or Cas A) – was accidentally discovered independently by two astronomers working at the same observatory in Chile, the Canadian Ian Shelton and the Chilean Oscar Duhalde (although Señor Duhalde was too busy at the time to tell anyone), and by the amateur Albert Jones of New Zealand. It was found quite accidentally (photographically by Shelton, visually by Jones and Duhalde) in the Large Magellanic Cloud (LMC) at 5th magnitude on the night of February 23–24, 1987. On a photograph taken the night before in Australia by Rob McNaught, it appeared at about magnitude 6.1. (Both Rob and Albert Jones tell their sides of the story in Chapter 5.) After its initial outburst, SN 1987A continued to brighten slowly but steadily until it peaked at visual magnitude 2.8 some 80 days later. (The visual magnitude of the LMC is 0.5.) Such behaviour was typical of the so-called Type II supernovae although some experts on such matters were surprised that it didn't become another one or two magnitudes brighter. A Type I would have become even brighter, perhaps as bright as magnitude -1.

After maximum, a supernova fades irregularly at first, but then settles down to a slow, steady decline of about a magnitude every two months. SN 1987A was still faintly visible to the naked eye nine months after outburst.

Like novae, supernovae of Type I, the offspring of white dwarfs with greatly more serious problems, occur where well-evolved stars are found, which is to say just about anywhere except where star formation is actively in progress. But Type II supernovae, the result of the collapse of super-massive stars, can be expected to appear in the dustiest and gassiest regions of galaxies. Why? Because supergiant stars shine so brightly; they burn up their hydrogen fuel voraciously. As a result thay can not last long. And so if we see supergiants, we usually always see mixed in with them the stuff from which stars are being made: interstellar gas and dust. Supergiants are the youngest stars we know.

These facts were initially deduced from where the two kinds of supernovae were found: Type I in all types of galaxies, but often away from the dusty, gassy spiral arms; and Type II only in spiral arms.

A warning: Type IIs will sometimes be difficult to detect, embedded as they often are in star- and nebula-rich regions, especially on deep photographs where bright spiral arms may saturate the film, making it difficult if not impossible to pick out even a bright newcomer. And the interstellar dust may further obscure the light, as it can and so often does in our own Galaxy.

Finally, to give yourself a decent chance to find a supernova, you should use a telescope or camera large enough, and in dark enough skies, to reach to at least 14th magnitude. Appendix D discusses magnitude limits of telescopes and cameras. Of the 63 supernovae of magnitude 15.0 or brighter that occurred in the 1980s, only three were brighter than 12 and only eleven brighter than 13. (See the bar graph, Figure F14 in Appendix F.)

4

Asteroids, variables, and other interesting phenomena

Tsutomu Seki. In the first chapter, we read that Brian Marsden singled out this man as the leading all-round amateur astronomer of the world when it comes to making discoveries. Besides comets, Seki has found (as of this writing) 22 asteroids (minor planets) and has picked up many more for which final orbits have not yet been calculated. How does he do it? We begin this chapter by considering asteroids and the various ways of finding them. Then we will go on to a discussion of other discoverables – variable stars of various types, meteor showers and meteorites, transient features appearing on or produced by the Sun and the major planets, and aurorae (borealis and australis).

Before beginning, we should record that Tsutomu Seki has discovered or co-discovered (with his close friend Kaoru Ikeya) six comets, one of them super-spectacular (Comet 1965 VIII; see Figure 2.3). In Chapter 5 we will find out more about this remarkable man and how he does it, and in his own words.

Part 1: Asteroids
Nomenclature

In naming minor planets astronomers use a system totally different to that used for comets. The first 24 letters of the English alphabet are assigned to consecutive half months of the year, so that, for example, the first minor planet discovered in the second half of February of 1989 was provisionally called 1989 DA. It was found at Palomar by Jeff Phinney. Some day he will have the privilege of bestowing a name on it – if the orbit of 1989 DA is well enough determined for the asteroid to be recovered at a minimum of two more oppositions. (This particular minor planet happens to be very small and nearly always very faint.) The next discovered asteroid in February was 1989 DB, the first in March, was 1989 EA, and so on.

Asteroids that have acquired well-established orbits are then assigned numbers and then later, names. The convention is to use both name and number, like (4) Vesta, (944) Hidalgo, (1814) Bach, (2309) Mr Spock, and (3935) Toatenmongakkai. For a while, there were logical gender conventions: well-behaved minor planets received feminine names; unusual asteroids masculine. But now with over 5000 numbered minor planets, any reasonable name is acceptable to the Commission of the International Astronomical Union that worries about such matters.

Orbits

The simplest way to characterise asteroid orbits is to divide them into two types: the normal and the unusual. Let us dispose of the generally less-interesting orbits first, those of the 'main belt' asteroids.

Perhaps 95 per cent of those approximately 4500 minor planets with well-determined orbits fall into the first class. They move with direct motion in nearly circular orbits with the value of *e* averaging 0.14, less than that of the planet Mercury. (See Appendix E for information on orbits.) Their average orbital radius is 2.7 AU or about 80 per cent greater than that of Mars; their periods of revolution about the Sun fall between 3.3 and 6.0 years, averaging 4.5 years; and their orbital inclinations rarely exceed 15°. At closest approach to the Earth, they are still somewhat more than an astronomical unit away from our planet, but, being at opposition, are in their full and therefore brightest phase.

And so these minor planets, measuring up to about 1000 km in diameter − (1) Ceres − bide their time orbiting the Sun between the orbits of Mars and Jupiter. Almost all main belt asteroids that get brighter than 15th magnitude have now been discovered, and since normal orbits are quite stable, don't expect to find any more brighter than this.

Now for the minor planets in unusual orbits. Most spectacular are the ones that cross inside the Earth's orbit giving them the potential of coming close to the Earth − the Earth-approachers and, sometimes, the Earth-threateners. Two (or three) groups are distinguished, although the division is perhaps arbitrary: the *Apollo* asteroids have their perihelion points inside the Earth's orbit, and the *Aten* group have periods less than one year. Obviously, all Atens also have their perihelia inside the Earth's orbit.

At the end of the 1980s, nine Atens and 63 Apollos were known, but since the hunt continues unabated, doubtlessly more will be discovered. (Eleven Apollos and two Atens were found in 1989.)

Another interesting group are the *Amors*: they come within Mars' orbit but stay outside of ours. A few can sidle up alarmingly close to the Earth at times: 1989 VB, discovered in October 1989, has a perihelion distance of 1.0052 AU and an inclination of 2.1°, which means that if its orbit remains unchanged, it has the potential of coming within a million kilometres of the Earth. With an orbital period of 2.52 years, 1989 VB has a close encounter with us every other time it goes around. Look out for it in November 1994 and then again in November 1999.

All the Apollos and Atens discovered in the 1980s appear in Table F6 in the Appendix, together with information on their orbits and such values as their brightest magnitudes at time of discovery.

On a few occasions, what were thought to be asteroids turned out to be comets. (2060) Chiron, for example, never comes closer to the Sun than 8.5 AU and consequently remains well frozen and asteroidal-appearing. However, professionals Karen Meech and Michael Belton have succeeded in detecting a faint coma around

Chiron proving that it is, in fact, a comet. When we bring up the topic of meteors, we will mention two other minor planets that may, in reality, be comets that have been reduced by solar heating to a rocky or gravelly core.

Brightness

Like any solid body without atmosphere, asteroid magnitudes can be predicted quite accurately: they follow the inverse square law of brightness. The largest departure from this simple behaviour arises from phase effects: for example, a 'full asteroid' brightens up because all shadows disappear as would be seen from Earth. The complete formula for the apparent magnitude of a minor planet appears in Appendix C.

Some asteroids are irregularly shaped, and many are shaded irregularly or spotted, so that when they rotate, they vary periodically in brightness. A few vary enough so that the fluctuations can be easily detected visually though a telescope.

Appearance

Only under the best conditions can the largest asteroids be resolved into tiny disks of light; even in a large telescope, an asteroid usually looks perfectly stellar. Normal minor planets at opposition drift by at a rate of about two-tenths of a degree a day, but an Amor, Apollo, or an Aten can whiz by much faster. (See Figure 4.1.) Minor planet 1989 FC, which came within about 700 000 km of the Earth in March of 1989, was zipping through the sky at over a degree and a half per hour. A bar graph, Figure F21 in the Appendix, shows how fast some other Apollos and Atens of the 1980s were moving.

1991 BA (see Figure 1.2) currently holds the record for known closest approach to the Earth: it missed us by 170 000 km, less than half the distance to the Moon. At magnitude 17, it could have been no larger than 10 m in diameter, about the size of a house. While experts disagree as to the details, had 1991 BA actually hit the Earth, it would have released the equivalent of about a megatonne of TNT and created a crater nearly a kilometre in diameter – or more likely, made a horrendous splash. To make a bad pun but a good mnemonic, an asteroid Aten (oughtn't) come so close.

As can be easily appreciated, virtually all minor planets are discovered photographically today – and in the direction of the sky where their phases are near full. A striking fact to keep firmly in mind is that an object with a surface like that of the Moon (and probably including most asteroids) is five magnitudes – *one hundred times* – brighter when full than when it is in a crescent phase equivalent to the three-day-old Moon. Obviously, the place to search is around the solar opposition point.

I will conclude this section on asteroids with some remarks adopted from an article written by ex-amateur Rob McNaught. (It appeared in the September 1989 issue of the magazine *The Astronomer* and is reproduced courtesy of Rob and editor Guy Hurst.)

Fig. 4.1. Minor planet No. 3757 = 1982 XB, as it appeared on a 46-cm Schmidt photograph taken by Eleanor Helin. This Amor reached 14th magnitude and moved across the sky at an angular speed of a minute of arc per minute of time.

In the first half of the 20th century minor planets were being discovered in such embarrassing profusion that the majority could be considered lost immediately after discovery! Observing techniques had gone far beyond the ability of astronomical computers (people) to cope with such huge amounts of data. To make matters worse, discoveries continued at an ever increasing pace as photographic emulsions improved and larger and wider field instruments were brought to the task. The situation was only relieved in recent decades by electronic computers which could store immense amounts of data and carry out searches rapidly.

When an object is reported to the Minor Planet Center, an initial check is made to identify it with already numbered, multi-apparition objects or with single-apparition objects with reasonably reliable orbits. If no identification can be made, the asteroid is given a provisional designation. (Occasionally, pressure of work at the Center causes a designation to be made before thorough checks can be carried out.) But the process continues in attempting to find identifications within the computer's huge data bank of observations that were made on only one or two nights. When a firm identification can be made with one or two such objects, an orbit can be derived and a general search for additional identifications is made, leading to an improved orbit. If during this process the current discovery apparition was the major factor in making the identifications, the object becomes known by that designation.

From the time of German amateur astronomer Johann Palisa's last discovery in 1924, minor planet discovery was solely in the province of the professional until the photographic discovery in 1978 of a minor planet by the Japanese amateur Takeshi Urata. This hiatus in amateur discoveries no doubt resulted in the perception among many that amateurs can't discover minor planets. This is also evidenced by the lack of treatment on minor planet discovery in books dedicated to amateur observational astronomy.

In a period of almost ten years from 1978 to mid-1987, amateurs discovered 197 minor planets (160 from Japan, 37 from Italy). Of these, 25 count as principal-apparition discoveries. Since then the rate has increased greatly.

Besides a wide-field telescope of modest aperture and dark skies, all that is needed to find asteroids is an up-to-date list of known minor planet positions, some know-how, and patience. Chapter 5 will include some wise words from the now oft-mentioned Rob McNaught, Tsutomu Seki, and Brian Manning. Here, too, you will find details of some of the discoveries made recently by Japanese and Italian expert amateurs.

Part 2: Variable stars
Types

The AAVSO, the American Association of Variable Star Observers, a misnomer of sorts since it is truly an international organisation, asks its members to monitor several thousand irregular variable stars, stars which never quite repeat their past

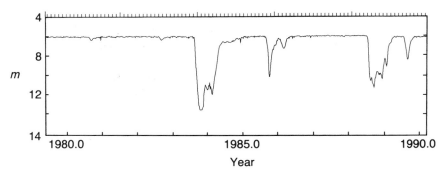

Fig. 4.2. R Coronae Borealis, the 'inverted nova' (AAUSO observations). Light variations are believed to be caused by clouds of fine carbon grains (graphite) ejected by the star itself.

behaviour. Many of these variables are the Miras, giant red stars that pulsate with periods ranging from around a hundred days to several years. Mira itself (Omicron Ceti) with an average cycle of 331 days can peak at anywhere from 1st to 5th magnitude and its period can be as short as 300 days or as long as a year.

Other pulsating variables include the Cepheids with periods from one to 50 days, RR Lyraes with periods under a day, and a host of other somewhat less common but well-studied types.

More spectacular and also regularly followed by amateurs are the dwarf novae like VY Aquarii, already mentioned, U Geminorum, and SS Cygni. They display a total rise in brightness of anywhere from a few to several magnitudes; SS Cygni, for example, brightens in a matter of hours by three or four magnitudes at intervals ranging from three to 13 weeks. All are probably binary star systems.

Distantly related to dwarf novae are the T Tauri stars which lie embedded in dusty regions of space and which vary almost continuously, partly because of the infall of material – a kind of interstellar rain – and partly because they are newly formed, and still unstable stars.

Just the reverse occurs with R Coronae Borealis and other stars of its class. Normally faintly visible to the naked eye at sixth magnitude, R CBr – or, as it is often called, R Cor Bor – may stay that way for years, and then suddenly, without warning, it fades for several weeks sometimes reaching as faint as 14th magnitude only to recover completely after many months or even years. (See Figure 4.2.) Obscuring clouds produced by the star itself are thought to be the cause of this uncommon behaviour, related to the more common eclipsing star systems.

Discovery

In Appendix F, Figure F22, appears a histogram showing the relative frequency of the different types of known variable stars. The Miras, usually exhibiting large magnitude changes and all very red in colour, are easily discovered and occupy first place in the census. Professional astronomers have discovered all Miras brighter than 12th or 13th magnitude, correct? No! Absolutely not. In fact it almost seems as if they gave up after a few years or so of photography.

Nowadays, new and often quite bright variable stars of all types, sometimes spectacularly variable, are being discovered by amateurs almost nightly. Nearly all these discoveries are made photographically: two sky photographs taken a night, a week, a month, or a year or more apart nearly always show a variable or two or more. Checking them against lists of known variables (see Appendix H) will, surprisingly often (embarrassingly often to the professionals), turn up a star not known previously to vary. Read the impressive failure-leads-to-success story of Dan Kaiser, and the sure-and-methodical procedures described by Mike Collins. They both appear in Chapter 5.

Part 3: Other phenomena
Meteor showers

In 1990 the valuable annual *Handbook* of the British Astronomical Association listed a score of annual meteor showers ranging in ZHR (zenithal hourly rates) from 75 for the well-known August-occurring Perseids and the December Geminids to a paltry 5 for lesser-known displays such as the Virginids, the Ursids, and the Alpha Capricornids. (In Appendix F, Table F7, we include a condensed version of this useful table from the BAA *Handbook*.) Meteor showers result from the passage of the Earth through the debris-littered wake of a comet, with the meteors themselves being the luminous result of the burn-up of tiny bits and pieces of detritus entering the atmosphere. Most incinerate at altitudes of around 100 km, but the big chunks can come lower – *much* lower (see next section).

The shower names indicate the approximate location of the *radiant*, the point in the sky away from which all the member meteors seem to move. This characteristic illusion produced by looking at a number of meteors all moving parallel to one another, becomes evident if one plots the paths of meteors on a star chart, especially if the shower is of short duration.

Not listed in the BAA *Handbook* is one of the most famous of all meteor showers, the Draconids, or Giacobinids, which was seen for the very first time in October of 1933, and since then at 13-year intervals. This shower was at its most intense in 1946 when the Earth ploughed through the cluttered orbit of Comet Giacobini–Zinner (1900 III) only four days after the comet itself had passed by. ZHRs as high as 2000 were reported in North America. The most recent display, on October 8, 1985, produced a substantial shower (ZHR = 200–300) over Japan.

Comet 'G–Z' passes perihelion every six and a half years meaning that at alternate visits to our neighbourhood, the Earth is at the other side of its orbit, safe from an encounter. The comet's debris, still confined to a narrow cylindrical zone around the orbit, produces a shower of short duration. Consequently, most of the activity takes place during a period of a few short hours. Watch for it the next time it comes by – in October of 1998.

A similar spectacle took place in November of 1733 when we encountered the debris of Comet Tempel–Tuttle (1699 II) now revolving about the Sun with a period of 32.9 years. Last seen in 1965 but expected again at the end of 1997, it has disintegrated to the point where comet material is strewn all along its orbit. As a result, *every* time we cross the comet's orbit (in mid-November) we are treated to a meteor shower, the Leonids. But at intervals near 33 years, the shower can achieve spectacular ZHRs; in 1967 estimates ran as high as 400 'falling stars' per hour. (See Figure 4.3.)

With the Leonids peaking again in November 1997 and the Draconids in October 1998, meteor observers have some exciting times coming up.

In recent years two apparent *asteroids* have been discovered that have orbits closely matching those of two meteor showers. 1983 TB now seems to be not an asteroid, but rather an old comet whose orbital debris causes the Geminids; and 1987 SY may be responsible for a weak shower known as the Delta Leonids.

Is there hope of finding new meteor showers? Absolutely, and for at least two reasons: first, there must exist hundreds of small undiscovered short-period comets whose orbits are continually being altered by close passages to Jupiter; second, faint long-period comets can remain undiscovered even though passing close to the Earth.

In mid-September, 1980, amateurs MacKinnon, Keen, and Kilardes noted a moderately intense shower (ZHR = 15–20) radiating from a point not far from Beta Cygni; the meteors apparently resulted from the close approach (0.034 AU) of Comet IRAS–Araki–Alcock in May 1983.

How, when, and where to look? Obviously, no more equipment is needed than a pair of good eyes, a dark sky, and a star chart on which to plot suspected shower members. And if you think you have discovered a new shower, you had better wake up someone else, experienced, who can confirm your conclusion. Put yourself in Brian Marsden's place: an isolated report from an individual whose name is not known to him carries little weight and hardly merits sending out an announcement to the world.

Better yet, take photographs; the location of meteor trails can be measured with precision and the radiant located precisely. And the images are *there*, in black and white (or colour). The best arrangement of all is to have two cameras separated by at least 25 or 30 km aimed at the same point high up in the atmosphere – like at about 100 km altitude. That is the region where most bright meteors light up. And although stationary cameras serve well, having the cameras following the stars makes it easier to see faint trails. Also, the location of the radiant can be determined without having to accurately time the meteor.

The best time of night to look is in the hour or two before dawn since at that time the Earth has rotated so that you, the observer, are situated on our planet's front side,

Fig. 4.3. The 1967 Leonid shower. Photo by Dennis Milon. Note the two meteors that occured almost precisely at the radiant located in the head of Leo.

the side facing in the direction towards which the Earth is moving. Like bugs plastered on the front of a car, meteors are swept up more efficiently on the front of the Earth. And finally, where to look? Anywhere, but overhead the sky is most transparent and the observing most comfortable. If you have determined where the radiant point is located, by looking in this direction you will see meteors coming nearly straight at you. They will, therefore, appear to move slowly outwards from this point; one coming exactly from the radiant will not appear to move at all, as Figure 4.3 shows.

Anyone with an interest in meteors should contact one of the several meteor societies (see Appendix I) and find out more about how to make reports on showers. These societies will also give you instructions as to what to do if you think you have witnessed a new or unrecognised shower. Obviously, independent confirmation is highly desirable, and it is wise to work out a joint observing programme with someone who will observe at the same time but from a different location.

Fireballs and meteorites

Fireballs, or bolides, have been defined in various ways, but perhaps the simplest definition is: a meteor that is long remembered for its brightness. All startle the casual observer, some even cast shadows, and a very few are visible in broad daylight.

Meteors are not always fragile bits and pieces of comets. A sizeable fraction of those seen on nights when no known showers are occurring are probably, in truth, miniature asteroids that were, moments earlier, on a collision course with the Earth. And of course some are not so miniature.

Dozens of fireballs have been seen and reported, followed by the discovery of fallen chunks of rock or stones, but the true frequency of this accidental coincidence is not great. However, the total amount of meteoritic material to strike the Earth's surface daily has been estimated to be more than 100 tonnes, most of the big pieces falling into the ocean. Sensitive seismometers left on the Moon's surface by the Apollo astronauts have recorded numerous meteoritic impacts in the area. A number of years ago, the Smithsonian Astrophysical Observatory set up a network of special cameras to photograph some of these spectaculars, and after several years ended up with a large collection of fireball photographs but only two recoveries of meteoritic material. Still, astronomy buffs and other night prowlers should record the details of any awesomely bright meteor on the chance that it might have survived its passage through the atmosphere. As the residents of Wethersfield, Connecticut, know so well, it can happen. (See Figure 4.4.)

Finding a meteorite even when the vicinity of the fall is known, is tricky business since most meteorites appear to the eye like ordinary Earth rocks. However, they usually contain sizeable quantities of iron – some are almost pure iron – and standard metal detectors can be useful. Meteorite suspects should be taken or sent to a geology laboratory for inspection and verification.

Solar and planetary phenomena

Despite the extraordinary successes of the space programmes of the United States and the Soviet Union, and even though giant ground-based telescopes located on high mountain tops have the ability to see to the edge of the visible universe, an

Fig. 4.4. Only a few meteorites have caused damage to dwellings. This photograph taken by
Phil Dombrowski shows a recent occurrence: Wethersfield, Connecticut, November 8, 1982.

impressive amount of valuable information on the planets reaches us though the optical systems of modest-sized telescopes owned by non-professional astronomers. The reason: changes are continually taking place and there are few professional planetary patrols, ground-based or orbital, in operation.

Venus, Jupiter, and Saturn never appear quite the same way twice owing to their atmospheres whose cloudy belt structures change from hour to hour. A spectacular example occurred in September 1990 when a brilliant white spot appeared in the northern part of Saturn's equatorial zone, timed perfectly, it seemed, for the Hubble Space Telescope. On Mars winds and frosts can churn up dust or deposit a thin, white crust on the polar regions of that ruddy planet.

Watching for these changes is not unlike looking for weather changes on Earth – a kind of astrometeorology – and is somewhat out of the domain of astronomical discovery (and the purview of this book).

Several organisations, the Association of Lunar and Planetary Observers in the USA, the National Association of Planetary Observers in Australia, and several sections of the BAA, the British Astronomical Association (Terrestrial Planets, Jupiter, Saturn, and Asteroids and Remote Planets) have regular programmes of planetary patrol. See Appendix I for names and addresses.

Without question, the most active and changeable body of the solar system is the Sun, but because solar storms can seriously affect radio transmissions and alter satellite orbits, there exist well-organised programmes of Sun patrol, both from the Earth's surface and from satellites and space probes. Still, a number of organisations comprised of amateurs who are fascinated by our own continually changing star have programmes for recording and reporting potentially disruptive solar events. The Solar Division of the AAVSO is one of the more active groups which regularly communicates with the US National Oceanic and Atmospheric Administration, one official keeper of solar data. In Britain, the BAA Solar Section is similarly occupied with the goings-on of our nearest star.

Solar storms cause aurorae. However, not all storms affect the Earth since the charged particles blasted off the Sun – electrons and ions – can go charging off in other directions in space never interacting with our upper atmosphere. And so there are organisations set up to report and record displays of the Northern and Southern Lights. (See Appendix I.)

As is well known, aurorae occur most frequently a year or two after the time when sunspots are most numerous, but there is no correlation, as many believe, between auroral occurrences and the season: they are just as frequent in the summer as in the winter. Of course, people living in those parts of the Earth near the magnetic poles see aurorae most frequently, at least in the winter when their nights are dark.

Planet X

Careful analyses of the motions of Uranus, Neptune, Pluto, and the Voyager space probes have given clear indication that an object with a mass characteristic of a major planet is orbiting the Sun beyond Pluto – the tenth planet. It has been dubbed Planet X, the X standing either for an unknown quantity, or for the Roman numeral '10'. Unfortunately, there has been no clear concensus where it is located. However, predictions suggest that it could be as bright as 14th magnitude and therefore within reach of many amateur telescopes.

What else can be said about Planet X? Its rate of motion across the sky would certainly be less than that of Pluto which moves about 70 seconds of arc per day at opposition. Therefore, photographs taken on successive nights should readily show Planet X's diurnal motion. A third night's photo – or better, several – would confirm it. Some predictions suggest that the orbit is moderately highly inclined (like Pluto's) suggesting that it could be well off the ecliptic.

It is hoped that in the near future the several experts who have worked on this intriguing but complicated problem can get their immensely difficult acts together and come up with a single, reliable prediction for Planet X's coordinates. Of course, if they do, *they* no doubt will be the first ones to look for it. In the meantime, astrophotographers: stay alert. It might just turn up on your next pair of photographs.

5

The discoverers: in their own words

At the suggestion of Simon Mitton, my good friend at Cambridge University Press, I wrote to several of the world's more successful astro-discoverers and invited them to contribute a few lines, two or three paragraphs, or several pages to this *Guide*. Admittedly sceptical at first, I sent off letters to some that I knew personally and to a few others from whom I especially hoped to hear, but my feeling was that many might want to guard their secrets, or perhaps just would not want to be bothered with what for many is a tiresome task of putting words on paper.

How wrong I was! The response was virtually 100 per cent. Given *carte blanche* to write whatever they wanted, the discoverers, mostly amateurs but in a few cases, professionals – or professionals 'moonlighting' (such a word!) – mailed back revealing and always interesting descriptions of their procedures, often including amusing accounts of their 'first', or other anecdotes. Many sent photographs showing them at work or with their equipment, and most provided short autobiographical sketches. Consequently, I asked others if they, too, would contribute good words of advice and describe some of their adventures.

And so in this chapter you will be able to read the actual words written by 20 successful discoverers. So that you can become somewhat personally acquainted with the discoverers, editing has been kept to a minimum. Except where clearly noted, none of the contributions have appeared elsewhere. My occasional intrusions and (I hope) clarifying remarks are always enclosed in square brackets.

The order of contributors is strictly alphabetical and at the end of the chapter appear short biographical notes and other remarks about the authors.

UNTITLED REMARKS
George E. D. Alcock, Peterborough, England

My interest began in meteorology, probably with the disaster which occurred at Louth, Lincolnshire. It was a brilliant cloudless day at my home in Peterborough [53 miles to the south], but a violent thunderstorm took place over the Lincolnshire Wolds. The torrential rain – a cloudburst – caused a flash flood which swept through the town causing £100 000 damage and drowned cattle, sheep, and people in their homes. I was 8 years old at the time.

The next year, 1921, was one of the driest years ever recorded and the River Nene at Peterborough dried up, and for the only time in my life we walked along the dry bed of the river.

About this time – during the summer of 1920, 1921, or 1922 – there was a very large partial eclipse of the Sun when it was high in the sky. Equipped with smoked glass fragments, my class was allowed to watch the event in the school playground.

On July 9–10, 1923, there was a spectacular all-night series of thunderstorms over London and all of SE England. As dusk fell a 'forest' of forked lightning could be seen approaching from the south. Among many buildings struck were the spires of two neighbouring 13th century churches and a collection of fine grandfather case clocks were magnetised by the electrical storms.

As a boy these events kindled my interest both in astronomy and meteorology. My father had a small brass $1\frac{1}{4}$-inch telescope with which I first observed the movements of Uranus near 44 Pisces. Later I could see the planet with the naked eye which demonstrates how dark the sky was in those days.

Some time around 1926 my Sunday School teacher allowed me to use a 2-inch hand-held refractor with a suitable very dark glass, and with this instrument I was able to see the double sunspot maximum of 1926–28, my first real astronomical observations. I obtained an early edition of 'Norton's Star Atlas' and rapidly began to learn the appearance of the night sky. Around 1929 I was allowed the use of a 4-inch Cooke refractor in a garden observatory some distance from where I lived and made my first sketches of lunar craters. In 1933 I made a sketch of the white spot on Saturn discovered by Will Hay [both a famous British comedian and an avid amateur astronomer]. It is the only sketch that remains from those early days.

However on the evening of December 30, 1930, came an event which changed my life. As I was walking over the old river bridge at Peterborough, a magnificent fireball was seen dropping down in the western sky on a track NE/SW. Looking up the address of the Director of the Meteor Section of the BAA given in 'Norton's Star Atlas', I wrote to him and eventually received a reply from I. P. M. Prentice asking me to join his observing team. Later that year (late June) I was asked to attend a meteor section meeting in the library of Sion College, Victoria Embankment, London. Among those attending were Collinson, Prentice, and Alphonso King. The first matter on the agenda was the announcement of the death of W. F. Denning who had died a few weeks before on June 7, 1931, and whose work from 1870 to the 1920s was acknowledged by all present standing in silence. Alphonso King took over the computing side of our meteor work and remained in charge until his sudden death in 1936.

Also present at that first meeting was an elderly lady, a Miss Grace Cook, probably already in her nineties, who had been an active observer under Denning and the one who had introduced meteor work to I. P. M. Prentice. She had never seen me before, and most likely had never heard of my existence until then, yet when I stood next to her in a quiet corner of the library, she whispered to me that she was sure I was the man who would take the place of W. F. Denning. I told no one of her 'forecast' but how true it proved to be.

So from 1931 until 1954 my main activity was the observation of meteors under the directorship of Mr Prentice. Contrary to the present work in this field we were not specifically concerned with hourly rates of major showers and sporadic meteors but devoted our energies on trying to obtain orbits, speeds of entry, and heights of burning out of individual meteors entering the Earth's upper atmopsphere. To obtain sufficient data it was necessary to work for at least 4 hours or more each night. To do this a base line between at least two stations was used to obtain accordances, and at once we found weather conditions in Britain's fickle climate repeatedly minimised results. If one or the other station was clouded out, the night's work was almost useless.

It was then that I made attempts to forecast weather conditions for future nights at either station. An Express Mail System then existed whereby letters or cards could be posted around 0800 at Peterborough and replies received from Stowmarket by 1730 hrs. All mail was sent using passenger rail services. This system requiring an extra sixpenny stamp has long since been defunct, but visual written data had many advantages over the telephone. Recent developments in Electronic Mail (E-Mail) do much to improve (replace) this long-forgotten service.

Prentice and I showed that these sporadic meteors were members of the solar system moving around the Sun and not fragments which had entered the solar system from outer space, quite in contradiction to the results achieved by professional astronomers in the United States.

After World War II, radio astronomers began to track meteors by radar, and their results confirmed our earlier conclusions.

Once the radio astronomers were convinced they were picking up meteors on their screens night and day, work not influenced by weather conditions, I decided to consider abandoning meteor research after claiming a record meteor watch at Quadrantid maximum (see Table F7 in the Appendix) from 5:15 p.m. on the evening of January 3 until 6:55 a.m. the next freezing morning.

Early in 1951 I received a position from Dr Merton at Oxford of a comet in the dawn sky discovered by one of the Czech astronomers, Pajdusakova. On seeing this fine miniature object with its tail, I realised at once that if I had been looking, I could have found it.

However, I had only the use of a 4-inch refractor with a very restricted field, but the search was started on New Year's Day, 1953. As no comet had been discovered in this country since Denning's in 1894, I knew that it was likely to take a very long time, so in the autumn of 1955 I decided to add another 'string to my bow', a search for novae using binoculars. I had missed discovering Nova Herculis 1934 by being forced to stop observing at 1 a.m., but this had sparked an interest in this field. The 5th magnitude Nova Lacertae 1950 convinced me that it was possible to discover ones not bright enough to be obviously seen by the naked eye, but they could be detected easily by binoculars.

An article by de Vaucouleurs in the RAS journal *Observatory* showed there were different types of novae, fast and slow. After reading this, I was hooked on attempting to make a discovery. So in 1955 I began to memorise the visual appearance of the night sky in binoculars down to magnitude 8.

In 1959 I was able to get a wide-field 25 × 105 mm binocular-telescope previously used by the Afrika Corps under General Rommel in their desert campaigns. Within two months I discovered two comets in a week, quite a change of fortune, for in the previous five years I had missed three comets by a distance of less than 5 degees of arc.

It had taken five years to find a comet. Six years were spent trying to memorise the innumerable patterns among the galactic fields before I was confident enough to start serious patrolling for novae. This was begun in 1961, and the first success, in July 1967, was the immediate recognition of Nova Delphini, the most interesting of my discoveries. On Easter Monday morning of 1968 my second nova was found. Nova Delphini was still bright so for several days two novae could be seen by the naked eye. It was astonishing to me that two such objects, visible at the same time without optical aid and discovered by an Englishman in such inferior weather conditions, could attract here such little interest. However, it proved that the many years, seemingly a lifetime, I had made to make these attempts was at last proving to be worthwhile.

I have now made 11 important discoveries, counting the outburst of recurrent nova RS Ophiuchi in 1985. This was seen after a long period of overcast skies, as was the discovery in 1983 of Comet IRAS–Araki–Alcock which passed closer to the Earth than any other comet since 1770.

After the discovery of the size of the nucleus of Halley's Comet in 1985, the radio astronomers at Arecibo (Puerto Rico) looked again at their data of the 1983 comet. Finding that a radar reflection had been recorded, they calculated that the nucleus of Comet IRAS–Araki–Alcock was of a similar size to that of Halley's – about 15 by 9 km.

My Comet 1963 III was one that showed a remarkable outburst of brightness at the end of May from $7\frac{1}{2}$ mag. to at least 4, whereas my second comet of 1959, although having a conspicuous tail as it approached the Sun, failed to survive perihelion passage. All these are interesting matters to any discoverer, but he is left to find out such data himself.

Nova Herculis 1991 apparently has proved to be the fastest nova on record.

As far as I am concerned I have been intrigued and thrilled to read about the many huge telescopes and complicated equipment that have been used in following up this recent discovery. (It's a pity I haven't been able to see many photographs of them from such places as Siding Spring Mountain, La Silla, Tucson, Palomar, Lick, Hawaii, and the observatory on the Canary Islands.)

Of course, besides making discoveries I have had several frustrating narrow misses. The tail of Comet Pereyra 1963 V was actually seen stretching almost horizontally across the eastern horizon and was mistaken for the headlight of a car travelling along the distant Norfolk ridge.

Another comet actually seen was Comet Rudnicki 1967 II. On a brilliant night

Fig. 5.1. UK comet expert Jonathan Shanklin (l.) and comet discoverer Roy Panther (ctr.) listen to wise words from George Alcock (r.) at a 1990 BAA meeting.

with a superb sky a diffuse very faint large patch was sighted near Gamma Ceti. After checking a star chart using a dim red light, I tried for half an hour to confirm its position but failed to see it again and so rejected the observation. After this disappointment and noting the tremendous increase in artificial light to the north-west of the site, I practically stopped all attempts at comet sweeping.

Several misses in the nova field, especially the bright Nova Cygni 1975, almost brought this work to an end. One soon discovers there is very little justice in this type of observing, and unless you are a committed 'crank', I would never suggest to anyone that they should take up the search for comets or novae in these adverse weather conditions.

[For more information on Mr Alcock's activities, see Guy Hurst's excellent article in *The Astronomer*, Vol. 27, page 252, 1991.]

MY COMET-HUNTING CAREER
Rodney R. D. Austin, New Plymouth, New Zealand

It has long been my contention that discovering a comet is really easy. The difficult part is finding one to discover. I have been living proof of this for some years. My comet-hunting actually started fairly soon after I accidently 'discovered' Comet Honda 1968c about two months too late.

In order to be successful, one requires access to good recent information regarding newly discovered comets, good charts, excellent instrumentation, and above all – patience. The latter is possibly the most important attribute of a prospective comet-hunter, apart from a good sense of humour. When your greatest competition discovers a comet just where you would have been looking on the same night, but for cloud and other distractions, a sense of humour is essential. I have nearly lost count of the numbers of the 'if only' breed of comets that I have missed over the years.

For example, take the evening in July 1975 when I was resident observer/technician at Mt John University Observatory [on the South Island of New Zealand]. Both telescopes were out of action, so the observer scheduled to use one of them, Gerry Gilmore, gave up in disgust. Naturally, having nothing I could do, I chose to go comet-hunting. The fatal distraction took the form of Gerry insisting on a night on the tiles in the bustling metropolis of Lake Tekapo (pop. approx. 300). Sure enough that was the night before [8th magnitude] Comet Kobayashi–Berger–Milon was discovered right in the middle of the area I had chosen to search.

Lesson No. 1: *Perseverance in the face of invitations to the pub is also necessary.*

More recently, I spent a fruitless half hour in 1987 checking the position of a definite comet (I could see it moving), before checking the ephemerides of known periodic comets which I always carry with me.

Lesson No. 2: *There is little fame or future in rediscovering Comet Encke. Always check the ephemerides.*

Of recent years, now that I am based in New Plymouth [pop. 46 000] and working for the local newspaper, I have had a succession of observing sites all about 15 to 30 km from the city. City lights are death on comets, as are the lights of the Japanese and Taiwanese fishing and squid boats that infest our coastline during March and April. On one night, my first site outside New Plymouth gave me a perfect view of more than 85 such boats, each generating around 10 megawatts of light. They lit up [nearby 2517-metre] Mt Egmont and turned the night sky into something approaching full moonlight.

By June of that year, 1982, the stage was set for the most momentous night of my life. After weeks of foul weather, the sky cleared briefly on the evening of June 18th. I did not get out to the site until about 2:45 a.m. The sky was marginal and a bit 'muddy', but at least the squid boats were gone. However, the wind was picking up. After about 45 minutes of battling the increasing haze and a gusty breeze, I decided that it just was not worth the effort, but started one last sweep up from the horizon.

At the top of the sweep at an altitude of about 42°, I noticed something very faint and fuzzy pass through the field. I quickly backtracked, and sure enough there was a large misty patch about 5 minutes of arc across. My immediate reaction was one of irritation because it meant checking out yet another obvious galaxy. My irritation abated rapidly on checking the *Becvar Atlas of the Heavens* and the *SAO Star Atlas*. Nothing was noted in the area, so I settled down to keep an eye on it to see if it was moving. But all that moved was the telescope as the wind got up even stronger and nearly blew the tripod over. So I packed up and went home.

At home I had a set of the *Canterbury Sky Atlas* which is actually a whole bunch of true photographs, and not just litho-printed from negatives. It required getting them out from the bottom of a box at the bottom of a pile of boxes, but the effort was worthwhile as the object was definitely not on the appropriate photograph. This left me with a little problem, as I live with my parents these days and the telephone is right outside their bedroom door. I did not really want to disturb them, so I merely left a message for them to wake me at 8 a.m., but without explaining the reason why. They must have thought that something had gone seriously wrong as they were most disturbed by the way I looked. Talk about a bundle of nerves! I sat up and squeaked, 'I think I've found a comet!'

Well, I most certainly had, and it turned on a very nice display in the northern hemisphere out of sight of New Zealand [4th magnitude with a 5° tail]. By waiting to report it when I knew positively that I had a comet could easily have cost me the discovery. As it was, due to a slight breakdown in communication, I did not get official ratification for another three days, by which time I really was a nervous wreck.

Lesson No. 3: *If you are positive, let nothing delay the report.*

My first discovery led to a slightly embarrassing incident. Some months before, my top teeth had been removed and I had delayed the fitting of a plate. With all the media interest, I decided that it was high time to do something about the mumbling. TVNZ came to interview me and my teeth having been installed the day before were not perfectly fitting. Half-way though the interview they started to clatter. I had to retire to remove them and then mumbled on as if nothing had happened. My friends all wondered why I sounded so different during the second half of the interview.

Lesson No. 4: *There must be a lesson there somewhere.*

By 1984 I had decided that the way I had built the telescope (a 6-inch reflector) made it too awkward to use, so I sat down and redesigned the whole concept of the mounting, although I retained the optics. The new system was constructed out of fine-grained particle board 12 mm thick. It was heavy (ultimately too heavy) but very stable, and the whole optical layout was in the form of a cross 4 with two optical flats folding the light through a right angle but not obstructing the light beam in any way. The instrument rotated in altitude around the optical axis of the eyepiece. This made for a much more compact telescope and made it very much darker internally. I completed it on a Sunday afternoon and test assembled it on the back lawn. I just couldn't resist projecting the Sun with it to see if there were any large sunspots. There were, so I called out the folks to see, and by the time they arrived the image had moved sufficiently for it to set fire to the inside of the telescope! After the resulting emergency action, the telescope was really none the worse for wear, but I have never used it for solar observing since.

The following Thursday night, I used the new telescope for the first time for comet-hunting, and got in about an hour before the cloud rolled over again. The sky did not clear for another three days, so on the Saturday night six days after the telescope was test assembled, I headed out into the country on a clear night.

I started observing Periodic Comet Clark which was on its second return since discovery at Mt John by Mike Clark in 1973. On that occasion I was comet-hunting only 100 metres away when he found the comet on photographs. This time I could just see the comet, and that observation put me behind my observing schedule, so I missed what became Periodic Comet Takamizawa three weeks later. Not that it particularly worried me because just two hours later I found Comet Austin 1984i on only the third occasion the rebuilt telescope had been assembled.

This time I had a different set of problems to contend with. Recognising that I had a comet was the least of my problems. It was bright and it was moving very quickly at about 20 minutes of arc per hour. The magnitude that I quoted to both Perth Observatory in Western Australia and Mt John Observatory that night was approximately 8. Wrong again! After all I could barely see it by naked eye! However, that original magnitude has gone down in the record books and has given rise to interesting speculation by various experts regarding magnitude surges. [Subsequent estimates gave the magnitude as 5.8.]

After solving the problem of how to dismantle the telescope in the dark in a hurry, I threw it higgledy-piggledy in the back of the car, and was off down the road to home with my head stuck out of the window due to an iced-up windscreen. When I burst into the house the dog came out screaming her head off, thinking about burglars I suppose. It took only about five minutes to get through to Perth Observatory – after receiving a stern reminder from the international tolls operator that, 'It is 2 a.m. in Perth. Do you really want to wake them up?' She must have thought me some kind of nutcase particularly when I informed her that friends enjoyed being woken at that time of the morning for the news I had. The ratification was much quicker on that occasion and I knew even before heading to work that night. [Rod works the 6:30 p.m.–2:30 a.m. shift.]

Comet 1984i developed a very nice antitail, but I never observed it. In fact I saw the comet on only three occasions. In one of my attempts to pick up the comet in the evening sky through holes in the heavy cloud which seems to be triggered by my discoveries, I inadvertantly picked up a faint misty cluster close to the track in the twilight. Wrong again! Only this time I telephoned the position to Perth Observatory causing great consternation and confusion for a while.

Lesson No. 5: *Keep your cool!*

Owing to job changes, first to the Black Birch transit circle, and then back to commercial printing in New Plymouth, I did not get back seriously to comet-hunting until mid-1987 when I 'rediscovered' Comet Encke. On this occasion, I missed my footing when dismantling the telescope and did a nasty injury to my back. It became obvious that the instrument was just too heavy to lift into place on the mounting, so I immediately started looking around for a replacement. This problem was solved in mid-1989 when I purchased a Meade 8-inch f/4 Schmidt-Newtonian through my good friends [and amateur astronomers] Barbara and Frank Ives of Auckland.

Another very good friend, Steve Ferguson built a new mounting similar to the one

Fig. 5.2. Rod Austin and his ingeniously mounted 20-cm Meade telescope.

for the 6-inch, and the telescope now rotates in altitude around the eyepiece. It performs brilliantly. The only problem is the fact that the telescope, having been designed for photography, has its focal plane about 13 cm outside the tube. I use a 40-mm Clave Plössl of 2-inch diameter with a 2× Barlow. Thus the exit pupil is kept small enough for efficiency. The magnification works out at 41× and the usable field is approximately 1.4°. The whole eyepiece assembly is very long and hangs out of the side of the tube by about 35 cm. It looks rather unwieldy but actually it works very well indeed. (See Figure 5.2.) The finder is a standard Telrad [see Appendix J] which I find very convenient to use for comets.

These days I do my observing from just outside [the inland town of] Inglewood. The new site has several advantages. First it is surrounded by trees, giving very good protection from the wind. It is way off the road so I do not get the problem of car lights. It is also quite a fair distance from Mt Egmont, so that I escape the sheet of cloud which rolls over the ridge, and wipes out the old site while the rest of the province remains clear. There are some disadvantages, as the horizon is higher than I would like.

On the night of December 6/7 [1989] I finally left work right on 2:30 a.m. to find that the night was still clear, as it had been for several nights. This time there was a small difference, however, as the strong southeasterly wind had died and there was no cloud out towards Inglewood. The real aim of the observing session was simply to check out Comet Okazaki–Levy–Rudenko. A previous observation had showed a broad condensation was evident. The tail could be seen extending to the diameter of the telescope field as a very faint spine of gas.

With only an hour to twilight, I finally decided that nothing would be lost by searching the direct southern sky from the altitude of the pole down to the horizon. After only half an hour, something extremely faint caught my attention. Flicking my eye around the field, I could see that there was something barely visible, forming an equilateral triangle with two stars of about 12th magnitude. I noted the time on my observing log as 14:52 UT.

Normally the position check is very quick, but this time I had a little difficulty as the field overlapped two charts of the AAVSO charts. Finally satisfied that I had the correct area, by checking with the 11×80 binoculars as well, I came to the conclusion that the object was real and that it was not marked on any chart I had with me. At this point I drew a field sketch, and made my one mistake. I took the right ascension on the AAVSO chart as being 3 min. less than the correct position. This normally would not matter particularly if I was merely checking off a known object, but this time it led to several extra telephone calls.

Lesson No. 6: *Always check, double check, and triple check. Everybody makes mistakes.*

On returning home I checked my field chart against the *Canterbury Sky Atlas*, and double-checked the BAA *Handbook* for periodic comets. I next rang Peter Birch at Perth Observatory. Having a friend interested in comets four time zones to the west has distinct advantages when trying to get a comet confirmed. Having passed on the wrong position, I rang Mike Clark at Mt John University Observatory, again with the wrong position. Luckily, I at least had the declination correct, so it would have been only a matter of time before the mistake was detected. I discovered the problem myself when I rechecked the position 'just to be sure', and that meant another telephone call to Perth Observatory. At Mt John, Alan Gilmore had already telexed the news to the Smithsonian Observatory, which showed a very gratifying faith in my observing ability.

At 10 a.m. NZ time, Peter rang from Perth and confirmed that the object was indeed moving, and that the third Comet Austin was on its way. Life has never been the same since.

[According to a letter received from Rod, the early predictions that Comet Austin $1989c_1$ might become 'the comet of the century' resulted in a flood of calls and letters from the media and from well-wishers.]

VISUAL NOVA SEARCH
Reverend Kenneth Beckmann, Lewiston, Michigan, USA

Lewiston, Michigan, a small community nestled among the jack pines and hardwoods of northern Michigan, is one of many communities in the area that boast of a Spring Morel Mushroom Festival. As an avid mushroom hunter, I enjoy hunting this delicacy. Especially in the north woods of Michigan, hunting morels demands a

Fig. 5.3. Reverend Kenneth Beckmann.

meticulous eye, an uncanny sense of direction, and a textbook knowledge of edible and poisonous mushrooms.

Hunting novae visually is not unlike morel mushroom hunting. Hunting for novae demands a meticulous eye, an uncanny sense of familiarity with the heavens, and a knowledge of the appearance of galactic novae.

For the last decade, I have searched from observing sites in the Ozark foothills of Missouri, to the highlands of northern Maine, to the cornfields of Indiana, and now among the northwoods of northern lower Michigan. Regardless where our travels have led us, I have enjoyed the challenge nova hunting provides.

My first attempts at hunting novae were chaotic and haphazard. I attempted to memorise the position of each star to the 6th apparent magnitude using *Norton's Star Atlas* and a pair of standard 7 × 35 mm binoculars. After a struggling summer, I adopted the AAVSO Nova Search Program, purchased a pair of 10 × 50 wide-angle binoculars and obtained a copy of the *Skalnate Pleso Atlas of the Heavens*. That following winter, I set about the task of studying the phenomena of galactic novae. Using Cecilia Payne-Gaposchkin's classic, *The Galactic Novae*, I prepared a paper, 'A Spirit of Search', about the distribution and frequency of novae which later appeared in the *Journal of the AAVSO* (1981, Vol. 10, No. 1). As a result of this study, I noted several areas of the heavens where novae appeared to congregate over the past one hundred years. Once I utilised the AAVSO Nova Search Program's search area location tables, I found both a convenient means for documenting my observations and for labelling areas where novae are more frequently discovered. The AAVSO refers to these areas as 'common' areas because they appear to produce a higher incidence of novae than others. Just as a morel mushroom hunter returns to his or her secret hideaways, so I return regularly to areas which the the AAVSO refers to as common.

Hunting novae encompasses an enjoyable pastime that I acquired from Manfred Dürkefälden, a West German AAVSO nova hunter. Dürkefälden familiarised himself with the heavens by piecing together a patchwork quilt of binocular star 'constellations'. Peter Collins, another AAVSO nova hunter, popularised this method by creating familiar binocular star patterns into a lion on roller skates, an extended branch on a swaying bamboo, etc., or whatever the imagination might conceive [see his contribution later in this chapter]. Using stars to the 8th magnitude, I too have designed and created a network of starry friends.

A technical understanding of utilising star atlases and optical equipment is also crucial. As I have mentioned earlier, I have graduated to a pair of 10 × 50 mm wide-angle binoculars. This type of optics appears to best support visual nova hunters. Its limiting magnitude is about the 9th magnitude or a little fainter. First, most visual nova searches are conducted to the 8th magnitude simply because the number of field stars below this limit is generally unmanageable for visual observers. Second, binoculars of this size are still reasonably light-weight and manoeuvrable. Arms do not tire easily and extended observing periods are possible.

While the AAVSO Nova Division has endorsed the *Skalnate Pleso Atlas of the Heavens* (out of print), Tirion's *Atlas 2000.0* is its contemporary counterpart. The field edition is recommended and is an excellent addition to a nova hunter's library. While it is unadvisable to use either of these atlases as a field resource for identifying binocular star constellations, both resources are excellent for verifying a possible nova. Don't forget an ephemeris. This hand tool dispels ghostly planets or asteroids which have been mistaken for novae all too often.

I continue to use a pair of 10 × 50s for most searches, although I have used a pair of World War II 3-inch binoculars for parts of the southern summer Milky Way. Although I use the Dürkefälden–Collins approach of familiarising myself with the heavens, I have also relied upon what I believe to be a photographic memory. I have adopted the binocular star constellation concept as a standard approach for conducting nova search. Using the AAVSO Nova Search Program as a reference for documenting my observations, I have familiarised myself with about fifty areas of the Milky Way. The majority of these areas are on either side of the galactic equator, running from the constellation of Lacerta to the constellations of Sagittarius and Scorpius. I observe or search for novae whenever clear skies prevail, first searching common areas and then additional areas in the Program.

While binocular star constellations, binoculars, and a suitable star atlas are prerequisites for a personal nova search programme, there are other considerations which are necessary to a successful programme of search. As I concluded in my article 'Spirit of Search', 'Properly done, nova search can be as personally exciting as it is scientifically valuable. It depends on one's commitment, patience and determination.' One continues to grow in the knowledge and wisdom of nova hunting and we are often compensated for our efforts.

With a meticulous eye, an uncanny sense of familiarity with the heavens, and a knowledge of the pattern of visitation of galactic nova, I have come to enjoy my secret hideaways among the star fields and star clouds of the Milky Way, where perhaps, like finding elusive morel mushrooms, hopefully I will one day discover yet another nova.

UNTITLED REMARKS
William A. Bradfield, Dernancourt, South Australia

Although I first became interested in astronomy at the age of 13 on my father's farm in New Zealand, my special interest in comets developed in 1970 at the age of 43. In the early part of 1970 I was greatly impressed with the splendour of the very bright Comet Bennett discovered by amateur comet-hunter Jack Bennett in South Africa. Then in August 1970 I purchased a 150-mm aperture f/5.5 refractor telescope for $60 from a friend in the Astronomical Society of South Australia who had assembled and used it for comet-hunting. I was inspired by Jack Bennett's discovery and with the procurement of this telescope became determined to find a new comet. Commencing on January 1, 1971, I searched for 260 hours until on March 12, 1972, I found Comet 1972f. I was very elated and became determined to find another comet, to prove to myself and others that my first discovery was a result of more than having some good fortune.

All comet enthusiasts who aspire to finding a new comet can be guided by the words of the late L. C. Peltier (discoverer or co-discoverer of 12 comets) of Ohio, United States of America, who said that the best advice he could give was, 'To find a comet, keep looking!' However, the chances of being successful are improved if the following procedures are undertaken:

1.   Conduct comet-hunting in country areas to obtain dark skies away from cities.
2.   Spend at least 100 (preferably 200) hours, spread throughout each year, sweeping both evening and morning skies.
3.   Make use of favourable weather opportunities as they arise.
4.   Form a search plan to gain a competitive advantage.

The need to comet-hunt in dark skies is probably the most important point, because comets are mostly faint telescopic objects. Ideally comet-hunters should aim to discover new comets as soon as they brighten to the limits of detection.

I live in Dernancourt, an outer suburb of Adelaide, the capital city of South Australia. Over the years I have travelled away from home to escape the light pollution of the Adelaide area and have set up my portable telescope on the side of little-used country roads or tracks. Currently, my journeys take 60 minutes to travel 70 kilometres. Fifteen years ago I was satisfied with observing sites which were not so distant from home, but since then the outer suburbs have expanded considerably and have added to the scattered light visible in the Adelaide night sky. Despite the extra time and expense involved, travelling further into the open country has given me an extra bonus, because interruptions to my comet-hunting by passers-by are now almost non-existent. In the early days I was sometimes interrupted by curious motorists (some of whom wanted to look through my telescope), occasionally by local farmers who thought I may have been there to steal their livestock, and on one occasion I was cornered by two policemen from a patrol car who thought I was stripping a stolen vehicle.

Since the arrival of new comets into the range of visual searches with portable telescopes cannot be predicted, the comet-hunter must assume that a new comet can be found at any time. Searches must be carried out (clear skies permitting) during the whole year, including the cold winter months.

Although I now use a 250-mm aperture Newtonian reflector for some of my comet-hunting, my main instrument is the 150-mm refractor. With this telescope I expect to spend 8 hours each lunar cycle covering the areas within 90 degrees elongation from the Sun, to give about 100 hours for the year. Although some comet-hunters have found a comet with less than 100 hours of searching, I believe that 100 hours per year is the minimum effort required by a comet-hunter who hopes eventually to find more than one comet. In my early days of comet-hunting, I was spending more time, typically up to 200 hours per year, with additional time being spent during the clearer skies of summer.

The records of comet discoveries in the past all show that more comets are discovered in the morning sky than in the evening sky, roughly in the ratio of 3 to 2, or perhaps 4 to 3. In the morning skies comets are sometimes discovered with a brightness much greater than the detection limit of small telescopes when they emerge from the sky area in the direction of the Sun. I have experienced the special excitement of finding two such comets, both at a total magnitude of about 5 and with tails 1 degree long.

Some amateur astronomers possess observatories in the country, away from the light of cities. However, unless one can live at or near the observatory, considerable travel to and fro is still a necessity.

Many comet-hunters like myself who do not possess observatories transport equipment to observe from country locations. From my experience, I believe there is a considerable advantage to the mobility provided by portable equipment. One can escape the unfavourable local weather features such as fog, low clouds, and strong gully winds that some sites occasionally experience at particular times of the year.

When I motor out from home, I have the option of selecting one of 9 different sites (3 for evening viewing exclusively, 3 for morning searches, and 3 for either morning or evening). If I find local conditions at the preferred site are unsuitable or are likely to deteriorate during the following 2 hours, I can drive to another site and conduct searches, usually without further complications.

A study of weather forecasts and a regular viewing of television weather maps and cloud-cover pictures help me to decide when to go out comet-hunting. Clear sky opportunities can sometimes suddenly occur without much indication from weather forecasts, so I try to make judgements on viewing opportunities by observing cloud movements and cloud types.

Unlike other areas of astronomical discovery, there is strong competition among comet-hunters. Consequently, as in the sporting arena, comet-hunters need to develop techniques and strategies to help them achieve their objective.

I have been aware of the success of Japanese comet-hunters since World War II; the names of Honda, Seki, and Ikeya are well known in comet-hunting circles. In my searches I have always regarded the Japanese as the leaders in comet discovery from the northern hemisphere. Whenever I go out to make my first searches at the beginning of a Moon-free period I search that part of the sky which can also be searched by northern hemisphere comet-hunters from a latitude of 35 degrees north. If the northerners have several days of cloudy weather, I might be able to search before they can and possibly make a discovery.

When I search the sky, I try to search as efficiently as possible. I avoid unnecessary overlap of sweeps and ensure that no strips of sky are missed in the overall area which is covered. I maintain that the horizontal method of sweeping is the best to cover sky areas which are closest to the Sun's position. Thus an alt-azimuth mounting or a variation of it is the most suitable arrangement for supporting the telescope. A unidirectional mode of sweeping gives the maximum sky coverage (without missing out strips of sky) in the time available. The most effective way of sweeping the sky above 30 degrees from the horizon is with a telescope on a mounting which incorporates a polar axis. If the telescope can also be positioned on the mounting so that during sweeps the changes of eyepiece height above the ground are limited, then the sweeping difficulties will be reduced. I have spent considerable time designing telescope arrangements which afford easy and comfortable sky sweeping. [See Figure 5.4.]

During my years of comet-hunting I have obtained considerable satisfaction from what I believe is a challenging area of amateur astronomer effort. I must confess that in considering the enormous extent of the universe, both in distance and time, I relate more readily to our nearest neighbours in space, the various bodies in our solar system. I know that spacecraft have journeyed across the solar system and expect that man will eventually travel beyond the Moon to Mars and the satellites orbiting the planets.

The passage of comets across the sky and their changing appearance is an exciting area of observation. To hunt for new comets and to find them gives me a special interest in the possibilities of new sky splendours which comets provide.

Fig. 5.4. William Bradfield, his wooden, adjustable-height mount, and his $60 (Australian) 150-mm refractor.

PHOTOGRAPHIC NOVA PATROLLING WITH TWO CAMERAS

Mike J. Collins, Sandy, Bedfordshire, England

Following my return from Australia to see SN 1987A, I decided it was time to do some serious astronomy and to get going on a nova patrol using 7 × 50 binoculars. This proved good fun from the point of view of learning a few constellation micropatterns to about mag. 8 but absolutely useless for spotting anything interesting. I remember one very cold evening lying on the (Sun?) lounger and thinking what I would say if a nova was discovered in this field the next day: I would only be able to say that I hadn't seen it – no record. After that I just had to start photographing and I haven't looked back since!

Most patrollers seem to model their technique on existing procedures. Martin Mobberley of Bury St Edmunds and Nick James of Chelmsford appear to use the technique used by Rob McNaught, whereas I was strongly influenced by Dave McAdam and his serendipitous discovery of faint [mag. 10.0] Nova And 1988. [See Table F2 in Appendix F.]

The cameras I use are Soviet Zenits – built like tanks; if anything breaks you should be able to weld it! They are very heavy cameras but rather sturdy and haven't let me down. The 135-mm lenses replace the original two Helios 58-mm lenses which are used in the stereo comparator. The two cameras are mounted on a Takahashi Sky Patrol 2015 which I imported from Japan. The cameras do not photograph the same patch of sky: one is aligned with the finder scope; the other points ten degrees away to the centre of the patrol areas below (or above) in declination. I use Kodak Tech Pan 2415 film and develop it in D-19 for 6 minutes at 20 °C.

I started out with the intention of trying to pick up faint novae and keeping to the original UK patrol fields, taking pairs of exposures of each field. [Mike writes that the UK Nova Patrol Areas are 121 regions of galactic plane down to declination − 30°. He covers the regions from 0° to + 79°.] That dictated the 135-mm lens which gives me a field approximately 10° × 14°. The Takahashi mount allows me to align the camera consistently with the longer dimension of each frame in the N–S direction and track accurately during the 14-minute exposures. I had studied which types of film were being used and what sort of exposures I was likely to need before choosing 2415 film, but the fine grain and red sensitivity, in my opinion, far outweigh the exposure duration restriction in terms of picking up the fainter objects. I should be using hypersensitised film, but I can't afford it. I've been warned that 2415 is liable to scratching and the less it's handled, the better, but to keep costs down, I buy 2415 in large reels and reload.

The red sensitivity should pick up on the H-alpha of a nova, but of course it also accentuates the maxima of the red Miras and semi-regular variables. It came as quite a surprise when I started finding variables listed in the GCVS [*General Catalogue of*

Variable Stars] with maximum [brightest] mags much fainter than the apparent limit of the film. Dave McAdam considered this to be a good sign and that I was doing something right! The recovery of NSV [New Suspected Variable] 1098 came soon after in January 1989, only a couple of months after starting photography.

I should explain how I inspect the film. I mount up each frame although I originally designed the comparator to accommodate strips of film six frames long. I decided the risk of scratching was too great and that the need for an archive of film was important if I was to look at the fainter objects. Mounting the film takes time but aids subsequent handling and storage. The comparator uses the two Helios lenses (optics expert Dave Frydman believes they're not radioactive) in a set-up which uses convergent direct stereopsis. The stereo effect achieved is equivalent to looking at a star map held about 25–30 cm below my eyes. The magnification is about 5. The film is illuminated by a long incandescent lamp which produces a warm, yellowish light. The mounted films are placed beneath the lenses on a stage of perspex [plexiglass] which is horizontal and runs freely left and right on wheels and also slides along the axles of the wheels. The wheelbase is about 30 cm long, enough to permit a slight twisting of the stage which also helps line up the two slides. I have a dominant left eye so I place tonight's slides to the left and the master on the right. The master sits over a hole in the perspex, and the left-hand slide is positioned over a much larger hole until stereo is achieved. The stage can then be rolled or slid to inspect the whole exposure without having to readjust for stereo.

How do I do the actual patrolling? In stereo if one eye spots something different, the brain actually perceives something 'wrong'. It's hard to explain the feeling, but usually it seems that the object is 'floating' just above or below the flat sky of the stereo. If the object is on both films but is of different magnitude, it appears to 'blink', and if the object has moved it appears to 'jump' [as the eyes try to accommodate the two images]. By briefly closing my right eye, I can tell if a suspect is on tonight's image. Moving the stage away from me alters the illumination angle of the negative and usually confirms any film flaws which appear bright in scattered light. If lighting aspect makes no apparent change, I bring in the witness shot [backup photograph] of the exposure which is displaced slightly in RA so that ghost images can be distinguished. If a suspect is bright/faint on both frames taken tonight, life becomes a little more interesting. By this stage I will have eliminated about 90 per cent of my queries as film flaws – I rarely find ghosts because I do not use a lens filter of any kind.

To check a suspect I first resort to my own maps of each patrol area which are 71 per cent reductions of the TVMPSA [*True Visual Magnitude Photographic Star Atlas*] with stars to mag. 8 highlighted and marked with the patrol boundary, constellation boundaries, the brighter variables, and details of magnitudes from AAVSO maps, charts, and any known sequences. In this way I can now eliminate approximately 60 per cent of remaining suspects at this stage. I then check with the AAVSO *Atlas* and if there's nothing there, it's the GCVS, the GCVS *Name Lists* [these appear at irregular intervals in the IBVS, the *Information Bulletin on Variable Stars*; see Appendix H], the

NSV, and other lists like Lennart Dahlmark's and Dan Kaiser's [these also appear in the IBVS] to check. If still I have not identified my suspect, I allow myself to get a little excited but insist on rechecking the GCVS and NSV because of their size. I obtain positions direct from the TVMPSA or *Atlas Stellarum*. In most cases I've actually been able to see my suspect on the TVMPSA copy which goes down to about mag. 12. If the suspect is not recorded on my previous exposures nor on the photographic atlases, I check the IRAS [Infrared Astronomical Satellite] *Point Source Catalogue*. Only a few objects get past this stage, and those that do get the adrenalin going!

The suspect could be a nova but more likely an asteroid that has come into the field – this is especially true if the suspect is at the edge of the patrol area. I have a [computer] program that I can run which gives me the positions of all asteroids brighter than mag. 12.5 at the time. Several asteroids have given me a nasty fright. At one stage I was only recalling those asteroids that could reach mag. 12, but following the discovery of a mag. 11.5 object on film I extended the search to 12.5, drastically increasing the number of objects on file!

I compare by using master negatives which may be several months old – I know I should be moving the master along in time so that I filter out the variables, but my repeat cycle is very long, and in any case I have found that having a high quality master is extremely effective when viewing stereoscopically. As a result, I'm likely to be slow to pick up on a nova, but in the meantime I am getting regular practice checking suspects, enlarging my collection of catalogues, and preparing sequences and finding charts. I calculate that each query that I send Guy Hurst [Editor of *The Astronomer*] must represent a minimum of 5 hours' work. In some cases my suspect has really been at the film limit or has shown very little variability, and so I always have to pinch myself and ask why, of the tens of thousands of stars that I've scanned tonight, should my brain pick on this one as being something different. In many cases it's not been until I've checked the IRAS *Point Source Catalogue*, or the *Dearborn Catalogue of Red Stars*, that I've been convinced of my suspect. Hopefully I'll be a bit quicker off the mark with a nova!

Because I am unable to cover the whole Milky Way in one night, I have a housekeeping program which lets me know which fields should be photographed tonight by giving me weights for each patrol area – so I do cheat! However, the weights do not heavily bias those areas which have provided past novae, so I like to think that I am being less biased than some patrollers [like W. Liller].

[WL: In my defence let me add that there are two competing philosophies here: The first says that one should patrol a fixed area – the Milky Way from Constellation A to Constellation B, everything north of Declination $-20°$, the whole sky – to be able to arrive at good statistics as to the frequency of novae. The other philosophy argues that the most important goal is to find as many novae as possible since no two are alike and, furthermore, one may just turn out to be a supernova. Both philosophies have their strong arguing points.]

UNTITLED REMARKS

Peter L. Collins, Boulder, Colorado, USA

My astronomical interest developed early, around age 6, with a fascination with meteor showers. Until college I was a quite casual amateur, observing all the major phenomena but having no programme of work as such. In college I was studying to be a professional astronomer, but gave that up to become a computer engineer. Shortly thereafter I became inspired by certain persons, most notably Dr Brian G. Marsden of the Central Bureau for Astronomical Telegrams. There was a certain 'gathering of the faithful' syndrome at the observing sessions for Comet Kohoutek on the Harvard Observatory roof which I found moving to see among professionals. I resolved to become one of the 'discoverers' in the amateur astronomical world. In those days (1974–1977) that meant comet discovery; so I began to undertake visual comet patrol with a 5-inch refractor and an 8-inch reflector. I had an ancillary interest in variable stars and thought that I might someday search for novae. It always seemed that the best way to do the latter was visually by means of memorisation, following G. E. D. Alcock, but I was unclear as to how that could be accomplished.

In April 1977 I moved to Tucson to work at the Multiple Mirror Telescope (MMT) and to pursue my amateur astronomy under the clear skies of Arizona. I continued comet and variable star work, and did some tentative nova work without memorisation using 11×80 binoculars. In the autumn of that year, I realised what was required was to memorise nova search fields. Stars seen in binoculars or a telescope can be memorised as naked-eye stars are – with constellations. I began in November 1977 to memorise the stars in Cygnus and Lacerta using 11×80 glasses in an urban environment. Within a year or so, I had completed all of the summer Milky Way and about half of the winter Milky Way. The limiting magnitude was 7.7 and the width of my field was 10 to 20 degrees on either side of the galactic equator.

It was a stroke of good fortune that I actually found a nova rather early in the game: in Cygnus on the 9th of September, 1978. Others, such as Alcock himself, waited quite a few years for their first success. It turned out that this nova was about a degree from the popular variable SS Cygni, so there were multiple independent discoveries. One, by Warren Morrison of Peterborough, Ontario, came six hours before mine, which fact I learned two days later. I would be lying to say that there wasn't some disappointment in that – but overall it was a very positive event and confirmed the validity of what I was doing.

The discovery of Nova Cygni 1978 was made with 7×50 glasses and with a gibbous Moon nearby, but I was up on Mt Hopkins, a mile or so from the MMT site. The anomalous star in the field being searched was quite obvious and led me to check the suspect immediately. The checking process took about two hours, and entailed examining photographic charts, using the 8-inch reflector to obtain a semiaccurate position, and checking against the *General Catalogue of Variable Stars* and an ephemeris of bright minor planets. I was particularly proud of the position determination, accomplished by careful interpolation among several stars in the *SAO Star Catalogue*. This position was said to be more accurate than that initially obtained from a photographic plate taken at Harvard Obseratory's Oak Ridge Station. (In subsequent discoveries my positions were much less good and even held up to some criticism.) Once my checks were done, photographic confirmation was obtained from the Oak Ridge Station, and the Mt Hopkins 1.5-metre telescope just 100 feet away was used to take a confirming spectrogram.

I did go on to make two further discoveries: of Nova Vulpeculae 1984 #2 and Nova Vulpeculae 1987. In the latter case the first discoverer by about two hours was Rev. Kenneth Beckmann, chairman of the AAVSO Nova Search Committee. [See Rev. Beckmann's contribution in this chapter.]

The limiting magnitude of my search has deepened to about mag. 8.0 in most places and 8.5 and even 9.0 in some special regions. My hope is to find some fainter novae than are ordinarily picked up in visual searches. Nova Vulpeculae 1987 was 7.3 at discovery which is about as faint as any visual discovery made as part of a regular patrol. I hope to find some novae fainter than 8.0 such as those that abound near the galactic centre in Sagittarius. Beyond that, I have a number of other astronomical projects to occupy me; in particular, I have yet to find a new comet.

UPS AND DOWNS IN SUPERNOVA HUNTING IN THE 1980s
Reverend Robert Evans, Hazelbrook, NSW, Australia

Bill Liller's request for me to contribute to his book has given me an opportunity to tell some personal stories about supernova hunting, many of which have not appeared in print before, but which are an important part of the business of being human in the process of doing science.

Before relating these, however, the basic aspects of visual supernova hunting need to be stated briefly.

The purpose is to observe as many different galaxies as possible, on a monthly or two-weekly basis, with an adequate telescope and from a good site. Any new object which appears in the field and which is not a normal part of the scene should be explored thoroughly as it may be a supernova.

In order to maximise the possibilities of this search, speed can be an important factor, as this may enable the observer to cover more galaxies more regularly. However, speed should not be an excuse for carelessness or for sloppy observing.

With experience, the quickest way to cover a number of galaxies is by 'star-hopping'. A galaxy can be found in a few seconds and examined with high magnification for some more seconds (perhaps ten or twenty), before one moves on to the next galaxy.

Again, with experience and practice, the observer must learn to remember clearly not only exactly where each galaxy is, but also what each galaxy normally looks like

Fig. 5.5. Reverend Bob Evans and his 41-cm Meade telescope.

in the observer's own telescope, so that he or she does not need to stop repeatedly to check against charts or photographs, and thus lose a lot of time.

If any suspicious detail is seen, it must be checked carefully and quickly. If the observer is convinced the new object is a supernova, then there must be a competent observer friend who is both willing and able to verify the discovery immediately. This friend must be sufficiently well equipped to do this job properly, and sufficiently sceptical so that only real supernovae get through the verification net.

It is only after this kind of verification that contact should be made with the Central Bureau for Astronomical Telegrams [see Chapter 9] in order to report the discovery. The Bureau will need to know all the details about the discovered supernova.

The Bureau tends to believe the reports of people whom they know from experience to be reliable. Beginners especially, must expect to be asked to present evidence of a new discovery which is virtually fool-proof, and so should not be offended when they are not believed easily.

With each of my discoveries, and with some other supernovae that I have observed, there is a little human story attached, and sometimes more than one.

67

1980

The first supernova that I ever saw was SN 1980N in NGC 1316. Late in November of 1980 I was at a friend's star party, and observed this galaxy with his long-focussed 31-cm reflector. I remember noticing that there was something a bit strange about the star patterns near the galaxy, but it did not trouble me enough to make me pursue it any more. So I forgot about it.

The two weeks after that were spent on holidays in Sydney with no telescope and a full Moon. On several occasions while I was in Sydney, I had a very strong feeling that I should ring up Tom Cragg [an amateur at heart, a professional astronomer by trade] at Siding Spring Observatory [New South Wales] and ask him to look at NGC 1316. On several other occasions I have had premonitions like this.

But if I had rung him up, what would I have said? I could have said, 'Tom, I have got this funny feeling about this galaxy. Please have a look at it for me.' If he had asked for some stronger reason than a feeling – I had none. And that is not a very good way of doing science.

A few days after returning home, I had another fine night. Upon observing NGC 1316, the supernova was glaringly obvious at 12th magnitude and two minutes east of the galaxy. (Dec. 17.)

Tom verified my discovery quickly, but then found that a notice about the discovery of the supernova had been issued about a day earlier. So there would be no point in my trying to claim an official share in the discovery.

The story was slowly pieced together that Chilean professional astronomers [Maza *et al.*, University of Chile] had photographed the galaxy back in late November but had not examined their plate. Again, they had photographed the galaxy on December 15 and had found the supernova. This led, of course, to their looking at their earlier plate.

So a 'miss' the first time round helped to give me a bit of experience which would lead to more confidence later.

1981

My first actual discovery could have come to grief the same way. My wife and I spent February and March, 1981, on long-service leave at an ocean beach (Brooms Head) not far from where we then lived at Maclean on the far north coast of New South Wales. Our four daughters continued going to school by bus, and a retired minister relieved me in the parish for that time.

Incidently, he was far more competent than I was at church work, and rather put me in the shade. I later thought that getting him to relieve me for the two months was the best thing I ever did for the parish.

In early February I observed NGC 1532, along with many other galaxies, and I have a sneaking suspicion that I actually saw the supernova in the arms of this spiral, very faint and briefly, but it did not register with me properly. This was followed by two weeks of foul weather – not uncommon at that time of the year.

February 24 provided the observation when I saw the supernova clearly and rang Tom for confirmation. Tom observed the galaxy and the new star with his 31-cm telescope, and compared what he saw with the ESO Survey [a photographic atlas of the southern night sky]. Nothing brighter than 19th magnitude was visible on the Survey at the site of the new star.

I had a very preliminary version of Gregg Thompson's chart of that galaxy, but I had no photos of the galaxy. [Gregg is the first author of the 'Supernova Search Charts and Handbook'. See Appendix G.]

Tom was so wary of making a mistake with the first discovery and thus creating a bad reputation for our search programme, that he waited for several days, observed the galaxy again and saw the supernova had varied in brightness. Finally, he got one of the professionals to observe it also. Photometry with a one-metre telescope showed that this star was not normal.

Then he felt more confident about reporting the discovery to the Central Bureau. And SN 1981A became an amateur visual discovery (in fact, only the third in this class).

While Tom was still going through all this agony of mind, I had been looking at more galaxies. One of them was NGC 4536, on March 1. Here, just east of the centre of the galaxy, I saw a bright star (mag. 12) that I did not remember from previous observations. So I went to consult with Gregg's early chart of this galaxy as I did not have any photos of this galaxy either at that time.

I had already noticed, and written down, that the directions (north, west, etc.) on this preliminary chart were wrong, but somehow I forgot that. On the chart, just 'east' of the galaxy, was a faint star, and after much thought, I concluded that the new bright star was probably the same as the faint one on the chart.

About a week later, I drove up to Tom's place on Siding Spring Mountain to visit and observe, taking my battered 25-cm telescope with me.

Imagine my frustration and anger when I there found that the star in NGC 4536 was SN 1981B, and I had seen it one day before it was found by a professional in the Soviet Union.

But the scene was not all bad because, as Tom and I observed from the front driveway of his home on March 10, more drama was to unfold.

Tom and I both observed SN 1981A, and while he made a variable star estimate of its brightness, I wandered off to other objects. I knew that SN 1980N was no longer visible in my 25-cm, although Tom might still be able to see it, just. But I could not resist the lure of NGC 1316, and when I looked, there was another bright star present.

I called Tom's attention to it. His immediate 'knee-jerk' reaction was that it could not be true, or there was some mistake – two supernovae in the same galaxy in such quick succession. But, almost as quickly, Tom knew that it was indeed possible, and it had already happened a few times in the recent records of supernova observations.

Again we used Gregg's chart and Tom's previous observations of this galaxy. We got Steve Lee [another working amateur] to look at it with his 20-cm and to report back to us. Steve could see the star without any problem, but he thought that the new star was the same as another very faint star marked on the chart close nearby. Tom knew that this was not the case, as he could see both stars with the 31-cm.

On this occasion we were trying to contact the AAVSO and the Central Bureau within a few hours of the first sighting of the supernova instead of waiting for days.

Thus SN 1981D became my second discovery within fifteen days and might have been the third.

The Chilean professionals had been taking regular photos of their supernova, SN 1980N, including ones on March 1, 2, 3, and 4. But again there was the problem that they had not looked at the pictures. When the telegram about SN 1981D went out, they found that the new supernova showed on their four photos at magnitudes 20.5, 18.0, 15.5, and 15.0. I had found it at mag. 12.7 about a day or two before maximum brightness. Incidentally, this was the first classical Type I supernova to be found visually by an amateur. The other three were all Type IIs. [See Chapter 3 about types of supernovae.]

1982

For the next 25 months I did not see another supernova, although about October 1982 I had another premonition. This time I had a very strong feeling that a supernova would appear in NGC 1187. This feeling prompted me to rise about 2 a.m. after the Moon set and to stare at NGC 1187 for over an hour, but without being able to see anything new. The supernova was found about ten days later, just after the full Moon, by an astronomer at ESO [the European Southern Observatory]. Actually, I never saw that supernova at all, even though I tried again later. In later years, this supernova was recognised as belonging to the new Type Ib class, although very few observations were made of it.

1983

1983 and 1984 provided a 'purple patch' [Australian for 'a short period of unusual success'] for me with four more discoveries in each year, bringing my list of finds to double figures in four years.

The second discovery in 1983 has proved to be one of the most important in my list so far.

I had just missed out on discovering a supernova in NGC 4699, having looked at the galaxy on the night when the new supernova was just rising past the threshold of visibility in my telescope. And this supernova, SN 1983K, was found very promptly by the Chilean group. Even if I had seen it, I may have been too late.

When I stumbled across the supernova three weeks later and discovered how I had missed it earlier, I felt very frustrated again. The next Sunday night after church, I went out with the irrational determination that I was going to find a supernova or else! After spending an hour working through the Virgo cluster of galaxies, I was just about to give up when I turned to M83, and there was a supernova 'loud and clear'. It was magnitude 12 and was instantly obvious.

It was cloudy at Tom's place, so Gregg Thompson in Brisbane confirmed it and reported the discovery to the Central Bureau.

Dan Green, the assistant at the Bureau, was on duty at the time. (It was Sunday in Boston.) He decided that he wanted more confirmation. After all, why had it not been possible to get support at Coonabarabran [site of the Siding Spring Observatory]? The weather conditions had not been a part of Gregg's telegram. Dan asked several southern observatories for help, including Tom at the AAO [Anglo-Australian Observatory]. So by the next night, professional confirmations were available from both ESO and the AAO, as well as an observation by Jan Hers [another working amateur] in South Africa, arranged through the AAVSO.

It is possible that his request for more confirmation actually helped to produce the wide-ranging observations that were made of this supernova by many observers. Another factor was clearly that the supernova was very bright as supernovae go, and it was in a prominent galaxy which had produced several other supernovae in recent times.

Whatever the reasons may be, this supernova was observed at many different wavelengths right through maximum. The VLA [Very Large Array of radio telescopes in New Mexico] found it to be unlike anything they had observed before, and it was the first Type I supernova to be 'seen in the radio.'

It was nearly two weeks before maximum light was reached, and this supernova, SN 1983N, has become one of the two main prototypes of a new class of supernova, now called Type Ib.

The third and fourth discoveries for 1983, in NGC 1448 and NGC 1365, were much fainter and harder to find, and required a bit of hard staring at the galaxy before I managed to see them. I have often thought of the many times I have glanced quickly at a galaxy without taking such a concentrated look, and I have wondered what I might have missed as a result.

As time went by, more aspects of the story came to light about the supernova in NGC 1365, SN 1983V.

I found it on November 25, a little before maximum light. Tom Cragg confirmed it and went up to the AAT [Anglo-Australian Telescope] where Michael Dopita was trying to observe objects in the LMC [Large Magellanic Cloud]. The area near the LMC was too cloudy, so the spectrograph was used on the new supernova, providing a kind of supernova spectrum that they had not seen before.

It was another example of very quick reaction to a new discovery which is possible with a large telescope sometimes, but not always, and which is more likely to happen if the professional observer is also keen to observe the supernova.

On the 28th, Professor [P. O.] Lindblad [of Stockholm Observatory] was observing the galaxy at ESO and discovered the supernova. He was excited about it and got several other observers to check on it. He was, I think, a little bit disappointed when he learned next day that he was not the discoverer of the supernova after all.

This supernova was another of the new Type Ib variety and was observed surprisingly little in the end. But this was a fate which befell many supernovae at that time and which is no longer so true now, I am glad to say.

1984

SN 1984E was my first discovery of this year, in NGC 3169. It was also the faintest that I found with the 25-cm, at mag. 15.1. It needs hardly to be said that the transparency that night was superb. Although it was so faint, the supernova was relatively easy to see. The little town of Maclean had several good nights at that time, and I could see the supernova for about three weeks before it got too faint.

The month of March, 1984, had been spent on holidays and was almost totally lost to cloudy weather on the coast near Maclean. So for one night, I drove 200 miles west, observed about 570 galaxies in the ten hours of darkness [almost one a minute!] and then drove back home. That is still my record for one night, and not in winter either. But, for all those observations, there was no supernova to see.

Three weeks later, in early April, SN 1984E turned up. Three days before I found it, a Russian professional and a Japanese amateur had photographed the galaxy. But, for one reason or another, they did not report their finds to the Bureau until after my discovery had been confirmed by Tom, and reported, and the telegram had been sent out with one discoverer's name on it. Because they reported their finds before the IAU *Circular* was sent out (and therefore presumably had no prior knowledge of my find), they were included as independent discoverers.

Even before the telegram went out, I reported my find to friends at McDonald Observatory [Texas], and Martin Baskell was able to discover and make observations of the very unusual hydrogen-alpha spike which was visible in the spectrum of this supernova for about a month. Michael Dopita was also able to observe it a few days later with the AAT. Such an enormous spike in the spectrum had never been seen before and has only been seen since once or twice. Mike Dopita's observations led to the first occasion that my name appeared as a joint author of a research article in a top professional astronomical journal. This provided some light-hearted comments from some other professional friends who looked down the list of authors and addresses. They all came from well-known universities and observatories except for me. My 'institution' was a church.

The other 1984 discoveries – three in 42 days – between July 20 and the end of August, contain a few other little stories.

There had been a lot of bad weather, and the full Moon was just past. On July 20 the night was fine. I knew it would probably be the only fine night for a while and that the Moon would rise about 10 p.m. So I worked frantically for four hours, observing 337 galaxies in that time, from Coma down through Virgo, Centaurus, Pavo, etc. The 338th was NGC 7184 in Aquarius, and there I found a supernova. Tom confirmed it quickly, before the Moon got too bright, and in the moonlight, a UK Schmidt [United Kingdom Schmidt telescope at the AAO] blue plate was exposed briefly. The Moon made a mess of the plate, but the supernova was visible on it. I had contact at that time with Robert Thicksten, superintendent on Palomar Mountain. He arranged a beautiful plate of galaxy and supernova with the 60-inch telescope, and Dr [J. B.] Bev Oke made one spectrum with the 200-inch [at Palomar Observatory]. One other partial spectrum was made with the AAT, and a few photometric measurements were made at McDonald Observatory.

The first problem I had was that the only report which went in to the Central Bureau was the initial discovery report. Nobody did any follow-up observations in such a way as to report the results of them to the Bureau.

As a result the supernova was not given an official designation. After a period, questions began to arise as to whether what had been announced in the *Circulars* was a supernova or not. After all, there was another star not far away from the supernova which showed on the Palomar Survey [photographic atlas].

This problem was finally cleared up, so far as the IAU *Circulars* were concerned, by several English amateurs who reported to the Bureau concerning photos of the supernova that they had taken. Also, Dr Ann Savage sent in a report from Edinburgh on measurements from the Schmidt plate. So the designation given to the supernova, in the end, SN 1984N, was not in line with the chronological order of the discoveries.

The second problem arose a few years later, when, for other reasons, efforts were made to determine what type of a supernova SN 1984N really was.

I obtained a copy of the AAT spectrum because, after two years, the results of unpublished observations made with the AAT become 'in the public domain'. This only covered the blue part of the spectrum and did not show any details which could help to identify it with any known supernova type.

Bev Oke also kindly provided us with a copy of his spectrum. In the beginning, the supernova had been classed simply as 'Type I' because Dr Oke had said there was no sign of hydrogen in the spectrum. At this later stage, he thought it might be Type II, although there was nothing at hydrogen alpha.

This difficulty had arisen because his spectrum (the only complete optical spectrum of this supernova) did not show any of the lines which would normally class it as a Type Ia or a Type Ib, or as a Type II.

Alex Fillipenko [University of California, Berkeley] also thought it might be Type II, but Sidney van den Bergh [Dominion Astrophysical Observatory, Victoria, BC] disagreed. Professor Craig Wheeler's [University of Texas] verdict was that the supernova was 'strange'.

Since 1984, a number of other strange supernovae have been observed more fully than SN 1984N. But this chance to learn from a strange supernova was largely lost through lack of observations.

At this point in the story we will return to July, 1984. After discovering what later became known as SN 1984N, there were eight days of rain. On the first fine night after that, I found a supernova of magnitude 13 near the centre of NGC 1559.

Because of bad weather at Coonabarabran and in Brisbane, no confirmation was possible in Australia. So I called up [amateur astronomer] Danie Overbeek, in South Africa. His telescope was not suitable for galaxy observing, but he reported back to say that he could just see the two stars in the galaxy, as I had explained it to him (one star is a normal foreground star). But he did not have any pictures of the galaxy to use for comparison, and he could hardly see the galaxy itself.

Because the galaxy was too far south to be seen from anywhere in the north, I had a problem knowing what I should do to confirm the discovery quickly. [You should have called me! WL.] So I decided to ring the Bureau directly and explain everything to Dr Brian Marsden. Gregg Thompson had also managed to see the supernova but had reported an offset from the nucleus quite different from mine.

Brian was bold enough to accept my report and announce the discovery which became known as SN 1984J.

I had a special friend among the professionals in the form of Dr Ron Buta who was in Australia at that time. He was having an observing run at Siding Spring at the time when this supernova was found. First of all, Ron got the observer on the AAT to use his spectrograph on the supernova. When they looked at the galaxy with the 3.9-m telescope, they picked the wrong star for the supernova because the other star was almost in the centre of the galaxy and could have been the nucleus. After failing with the first star, the observer on the AAT gave the supernova up as a bad job, and went back to his own observing project.

Ron was observing himself with a smaller telescope (60 cm) and was using a photometer. With this he measured the colours of the star in the centre of the galaxy and was able to compare his results with photometric colours of the nucleus which had been published some years before. In this ingenious way, he managed to prove that the star in the middle of the galaxy was a supernova without using a spectrograph.

A UK Schmidt plate was also taken of the supernova, but the high surface brightness of the galaxy, combined with the blue-sensitive emulsion of the plate made the star in the middle less clear in the photograph than it appeared visually.

Peter Anderson, an amateur friend who took pictures of all of the supernovae using Tri-X film with his 41-cm telescope, produced pictures of this supernova which made it stand out very clearly.

The end of August brought a supernova in NGC 991 which became known as SN 1984L. It was found just before first light [of dawn], and NGC 991 was the last galaxy that I planned to observe that morning. As a result, Tom Cragg could not confirm the discovery until the next night (which he did). But, in the meantime, I contacted

Robert Thicksten at Palomar again, and he was able to secure quick confirmation through the help of the observers who were using the 18-inch Schmidt camera. Later he had a photo taken with the 60-inch reflector.

Spectra were also taken promptly at McDonald Observatory and at several other places which resulted in a report in the IAU *Circulars* that the supernova had been found about a month after maximum light.

This I had to doubt and reported that the supernova was not visible at all when I had observed the galaxy a month earlier.

This led to the realisation that SN 1984L was one of these peculiar supernova with a strange spectrum like that of SN 1984N in M83.

In what were then called 'peculiar Type I', the characteristic [red] silicon line in the spectrum of a Type I supernova at maximum light is never present.

In classical Type Is, the silicon line is strong through the time of maximum light but disappears about two weeks after maximum. That is why they said SN 1984L was a month old when, in fact, it was not.

Craig Wheeler, especially, realised that because this supernova and the one in M83 were both Type Is (not having hydrogen lines in the spectrum), were the same as each other, but were different from nearly all the other Type Is (in not having the silicon line at maximum), then they must constitute a separate class of supernova altogether. This represented a breakthrough in supernova studies. The new class soon became known as Type Ib.

Because SN 1983N and SN 1984L were both fairly well observed, they became the prototype examples of the new class.

1985

After that, there followed 13 months without any new sightings. The CSIRO [Australia's government research organisation] awarded me a new 41-cm telescope, and the family moved from Maclean to Hazelbrook, in the Blue Mountains, some miles west of the city of Sydney.

1985 was marked by a trip to India, as guest of the International Astronomical Union, to give a short paper on my observing at a Joint Discussion on supernovae, at the IAU General Assembly.

This was, of course, a tremendous experience for me in a great many ways. It enabled me to meet a good many of the well-known and leading professional astronomers, and to meet others involved in the search for supernovae such as Dr L. Rosino from Italy, as well as those who gave the other papers at the Joint Discussion. My paper was published in the official proceedings.

It is not often that an amateur is asked to address such an occasion, and it was indeed a great honour for me.

Arising from this occasion came a request from Dr Sidney van den Bergh to join in publishing a study of the rates at which supernovae appear in nearby galaxies, based upon the records of my observations. The first version of this was published in 1987 with Dr Robert McClure as the other author. A revision was published in 1989. And this has become one of the standard studies of supernova rates with two authors from the Dominion Astrophysical Observatory and myself with my unusual 'parent institution'.

1985 also brought one supernova – SN 1985P in NGC 1433. The Blue Mountains region has a great deal of cloudy weather, and sometimes I drive west for 70 miles to get a clearer sky, near the town of Bathurst. Ron Buta had just left Australia to work for a while again in Texas, and NGC 1433 was his favourite galaxy. The pattern of its spiral arms were a classic example for him of a certain point which had featured in his [doctoral] thesis. So it gave me a lot of pleasure to find this supernova in 'his' galaxy, just as it gave him pleasure as well. Ron had helped a lot in the development of Gregg Thompson's supernova search charts, both by his encouragement, and by providing especially a number of photometric measurements of stars which appeared in the fields of some of the galaxies.

1986

I wanted very much to find the first supernova of this year because it would be the first supernova to be found with the new telescope, and from Hazelbrook. Similarly, the first supernova in 1981 had marked the start of discoveries with the other older telescope. It would also mark the start of a new five-year search effort.

For some years, also, the [University of California] Berkeley Automated Supernova Search had been getting their equipment, etc., together, and I expected to have competition from them before long. The Chilean search had terminated about the end of 1983.

So I was very pleased and excited when I picked up SN 1986A in NGC 3367 in Leo.

There were three bad false alarms in 1986 that got into the list of official designations before they were found to be false. Thankfully, I was responsible for only one of these. The other two were made by professionals. My false alarm was due to mistaking the clump of H II [ionised hydrogen] in NGC 5253 for a supernova with the new telescope. It was listed as SN 1986F.

But SN 1986G was NOT a false alarm. It was a 12th magnitude supernova (at discovery) in the famous dustlane of Centaurus A [NGC 5128]. What a place for a supernova to appear!

It was found over a week before maximum light and brightened to mag. 11.4.

An interesting side-light about this supernova concerned a colour photograph of Centaurus A which David Malin had taken with the AAT showing the supernova.

Six months previously, astronomer Ray Sharples had applied for a photograph of this galaxy in order to study the globular cluster system of NGC 5128. A night in late May, 1986, was allocated and it was agreed that the plates would only be taken if the seeing conditions were 'superb'. This was done because the star images had to be small enough to distinguish them from the fuzzy outlines of the tiny images of globular clusters which were at such great distances. (Based on these plates, Ray later identified 44 globulars using a fibre optics spectrograph.)

When the night arrived, the seeing was indeed superb, and David Malin exposed three black-and-white plates which were sensitive to different parts of the spectrum, enabling him to create a colour picture by using them together. Everything went well until it was discovered in the darkroom that the third plate was a failure. It was some time before it was possible to replace that plate, and by that time the supernova had disappeared. So the resulting colour photo which was published by the Anglo-Australian Observatory shows a marvellous picture of the galaxy, but the supernova is green.

The distance to Centaurus A had previously only been guessed at – anything from seven million light years up to twenty-six million.

The supernova enabled astronomers to refine this somewhat, and they now estimate the distance at about ten million light years. This makes NGC 5128 only a little more distant from us than the galaxies in the local group and in the Sculptor group [NGC 45, 55, 247, 300, etc.].

Some astronomers believed that only about 3 per cent of the light from the supernova managed to reach us through the material in the dustlane, because the supernova was some distance inside the galaxy – largely hidden from view. They estimated that if the supernova had not been obscured, it would have been five magnitudes brighter – about magnitude 7 – and almost visible to the naked eye.

It had previously been believed by many astronomers (but not by all) that classical Type I supernovae were all the same, having all been caused in identical situations and by identical explosion methods. As a result, they could easily be used as yardsticks for measuring great distances in the universe.

SN 1986G put the nails into the coffin of the view that this sort of thing could be done easily. This supernova had a life-cycle which was somewhat faster than the classical Type I prototype (SN 1981B), and it showed some other peculiarities as well. It became clear that these explosions were all different, would have to be understood much better, and would all have to be studied much more carefully before they could be used as cosmic yardsticks with confidence and accuracy.

Max Pettini was one member of a team which used high dispersion spectroscopy to observe gas clouds in the line of sight to this supernova and discovered several intergalactic gas clouds between ourselves and Centaurus A which were previously unknown, as well as being able to make a study of clouds in our own Galaxy and clouds in the dustlane of Cen A.

SN 1983N had been used in the same way, and a few of the other brighter supernovae. Although detectors are so much better now than they used to be, only the brighter supernovae can be studied in this way, at this stage at least.

The other discovery that year was 1986L in NGC 1559. This provided me with the only amateur instance (so far) of an observer having official recognition for finding two supernovae in the same galaxy. Several professionals have achieved this, but nobody has yet found three supernovae in the one galaxy.

1987

1987 was a tremendous year for supernova studies with which I had very little to do.

SN 1987A in the LMC appeared in late February. The night it appeared I had another premonition that something would happen. It was cloudy at Hazelbrook, and so I wanted to go west to my site near Bathurst to see if I could find anything. I rang a fellow minister at Bathurst to see what the weather was like out there and was told it was '50–50' – not really a good chance after a long drive. So I followed my wife's advice and went to bed.

Several days later when Rob McNaught rang me up to tell me all of his story [see contribution by Rob in this chapter], my response to his question as to whether I had heard the news was, 'No. What have I missed?' Because of the premonition and his call, I knew I had missed something.

However, several days later, I picked up SN 1987B in NGC 5850. This was a faint star near mag. 15.0. It had a strange spectrum which still defies interpretation, although it was well observed at McDonald Observatory at least.

I saw this supernova three days before the official discovery date, making a mental note to follow it up, asking Rob to check it, and then forgetting about it for a day or two. It had not been a sighting which instantly struck me as an obvious supernova. So when I remembered to check it and observed it again, it became clear that it was indeed a faint supernova, a long way out among the outskirts of the galaxy.

Then in December of 1987 my friend, Win Howard, a lawyer by trade, and I were again at Tom Cragg's place on the mountain. Again we observed from Tom's front driveway.

Again I had a strange feeling that a discovery might be made, and I even bought in advance some apple juice to celebrate the discovery that I hoped would be made.

The second evening was fine, and after checking a few galaxies with the old 25-cm telescope, I saw a supernova near the nucleus of NGC 7606. Within two hours we had checked the computers in the Central Bureau for asteroids, obtained a plate with the Uppsala Schmidt, by courtesy of the observers, and got exact positions from both the AAT and the 2.3-m telescope. A spectrum was obtained the next night. So again, the discovery of 1987N was confirmed, reported, and celebrated within a few hours of the discovery.

Among the professionals who observed this supernova quickly after the discovery was Paul Murdin who organised the British telescopes on the Canary Islands [the Teide and the Roque de los Muchachos Observatories] to observe it. The resulting notice of their work in the IAU *Circulars* represented the first published results of observations with the new 4.2-m William Herschel Telescope. The supernova was apparently in front of the spiral arm of NGC 7606 against which it appeared because they saw spectral lines caused by intervening gas in the Milky Way, but no lines from any interfering gas or dust in NGC 7606.

1988 and 1989

1988 produced a shared discovery in M58 [with Ikeya of Japan], and in 1989 I had a discovery in M66 which was one of the brightest supernovae [mag. 12.2] found in northern galaxies in the last few years.

But I am hoping that there will be another 'purple patch' before long which will not only produce a new crop of discoveries, but will also produce another group of little stories like the ones I have been telling to you here. It will also bring me new friends like the ones I have already met in the worlds of amateur and professional astronomy.

[End note: Since the beginning of the 1990s and up to publication date of this *Guide*, Reverend Bob has indeed had another purple patch, discovering SNe 1990K, 1990M, 1990W, and 1991X. His total as of October 1991: 22. He usually writes up each discovery and mails out a dozen or so copies of each narrative. *The Astronomer* in England and *Pulsar* in France frequently publish his remarks.]

UNTITLED REMARKS
Gordon Garradd, Tamworth, NSW, Australia

I've been searching the SMC and LMC [the Small and the Large Magellanic Clouds] since late September 1986, picking up LMC novae in September 1987, March 1988, October 1988, and January 1990. I'm sure there were novae in 1989; however, I was away bushwalking in Tasmania for the first 2 months (I walked over 400 km!) and the weather and other commitments restricted me to only two or three patrol photos per month on average. To have full coverage I think that I would need to photograph every 5 to 8 days, as some novae are very fast.

I use a 300-mm f/4.5 Nikon telephoto and an FE-2 body and hypersensitise my own Technical Pan 2415. Exposures are typically 8 to 12 minutes, depending on conditions, although I sometimes go for 20 minutes to go deeper after a break in patrolling. I take a second photo, usually 5–10 minutes for confirmation, as I find that there are quite a lot of emulsion faults (or maybe satellite flashes) that look exactly like stars. Generally my limiting magnitude is about 14.5, and if I look carefully I can pick up stars half a magnitude fainter than this.

My searching method involves placing two negatives of similar density on a light box. (Actually it's only a piece of milky perspex plastic resting on a couple of tins or slide boxes with a small fluorescent light underneath.) I then place two 8× eye loupes, one on each negative, and move them until the images are merged

Fig. 5.6. Gordon Garradd and friend.

(stereopsis). It took me only a small amount of practice to use this method successfully and I can regularly pick up a magnitude difference of 0.5 on variable stars. In fact I've found two new variables, both long period, that I can't find catalogued elsewhere.

My field is oriented along the bar of the LMC and along the long dimension of the SMC including 47 Tuc near the corner of the frame.

I always search the negs within 24 hours and have been fortunate to pick up most of the novae near maximum (according to Rob McNaught).

I have to drive out of town in winter to see the Magellanic Clouds as they become lost in the light pollution of Tamworth. I live on the NE edge of the city (pop. 33 000) so the sky overhead and from NNW to SE is reasonable. However, I prefer to observe with a dark sky and have five sites at various distances up to 80 km away where the sky is very dark. The best site in the mountains to the SE is the one most distant; it is at an elevation of 1300 metres and stars of mag. 7.0 are usually visible overhead. Unfortunately, I can't afford to drive that far too often and usually settle for a closer site, that is not quite so dark, but still much better than my backyard. I do visual comet hunting for an average of about 5 hours per month – but no success so far!

I gave up a clerical job in an accountant's office in 1984 so that I could pursue my favourite hobbies of astronomy, photography, and bushwalking. I make a living from some part-time accounting work (I can handle $\frac{1}{2}$ day a week behind a desk, but 5 days is just too much!). I also make some money from my photography, doing some E-6 colour processing, and also from royalties on sales through a London-based slide library. My main photographic interests are native subjects – wildlife, wilderness, and especially severe weather phenomena such as lightning and storms. My aim is to make enough income from royalties so that I don't need to do any other part-time work. At the moment I just make enough to live on, but I have plenty of time for my three hobbies.

P.S. I also build my own telescopes and mounts including the mirrors. My current instruments are a 20-cm f/9 Dobsonian and a 25-cm f/4.1 equatorial photographic instrument. I'm planning to start on a 48-cm f/5 Dobsonian shortly. [See Figure 5.6.]

UNTITLED REMARKS

Eleanor F. Helin, Jet Propulsion Laboratory, Pasadena, California, USA

The programme, 'Palomar Planet-crossing Asteroid Survey' (PCAS) was initiated by me in January 1973. The Shoemakers [see the contribution by Carolyn Shoemaker later in this chapter] started a separate programme using the same telescope in September 1982 [the 46-cm Palomar Schmidt].

A pair of observers working really hard for 12 hours may take 80 films with 4-minute exposures, but this is overkill as they cannot begin to keep up with any realtime scanning. Hence, when objects may be found days or weeks later, they are frequently lost due to the long delay.

Fig. 5.7. Eleanor Helin at the guide telescope of the 46-cm Palomar Schmidt.

I prefer to operate the telescope with a minimum of three people (sometimes four) to keep the telescope continuously in operation while concentrating on immediate inspection of results. With 6-minute exposures we take about 70 photos during a full night near new Moon in winter and 40–50 in the summer. I believe the Shoemakers have also long ago stopped 4-minute exposures which gave them nearly 80 (40 pairs) per night.

Kodak 4415 film (4 × 5-inch Technical Pan), a fast, fine-grain emulsion, is and has been used by us for several years. My exposure time is 6 minutes (with 8-hour hypersensitisation at 60 °C) to produce maximum magnitude with minimum background. The grain affects the optimum ability to detect faint, fast-moving objects in a reasonable length of scanning time. The 4-minute exposures lost several orders of magnitude. On the other hand, exposures of 8 minutes or longer can actually lose objects into the grain structure of the emulsion.

As for sky coverage, we are extremely limited especially when you consider objects with close-in orbits. We need to improve our ability to see/detect those important closest neighbours in space. It is frustrating to realise how difficult it is to detect a really close object.

When Eugene Shoemaker started his programme he selected only the optimum observing runs – the longest winter runs with best weather. He always preceded me at the 18-inch telescope. He also left the shorter, less promising runs during summer to address other activities. That was changed about 2 years ago. We now alternate on the telescope during dark runs, and he has since requested time throughout the year sending someone to carry out the runs particularly in the summer months or whenever he is not able to be present. Apollo 1989 FC was discovered by retired NASA scientist Henry Holt who preceded me at the 18-inch telescope in March 1989 during one of Shoemaker's absent runs.

I have continuously spent 5–10 nights (for 8 years I used both 18- and 48-inch Schmidts back-to-back) at Palomar each dark-run since January 1973. [See Figure 5.7.] Actually, in 1972 I spent time carrying out feasibility runs for what would become the longest-running survey of NEAs [Near Earth Asteroids].

The first week after I return to JPL after an observing run are extremely busy with discovery and follow-up astrometry to do, together with other commitments in my schedule. Even now, though clouded out (I am writing this at the Schmidt telescope waiting out a heavy cover of cloud), my time is spent using the stereomicroscope scanning films taken during the first two nights.

I discovered Apollo 1989 PB [see Chapter 4 and Appendix F] about two weeks before its closest approach to the Earth at about 3.4 million kilometres. I got Steve Ostro [of JPL] on it from Arecibo (the large radio telescope in Puerto Rico, part of the National Astronomy and Ionosphere Center), and the results produced the first two-dimensional images of a NEA.

ENQUIRING STARS, MY PATHETIC ASTRONOMY
Minoru Honda, Kurashiki-shi, Okayama, Japan

It seems to have all started in 1923 when I became fond of the celestial bodies, especially comets.

A boy who was born and raised in a humble mountain village in the Chugoku district in western Japan had no wealth but the natural environment. The night sky lined with high shadows of mountains was not vast, but was decorated with beautiful stars as well as the immense Galaxy which truly gave the pure, primitive view. Such a fantastic sight had absorbed my heart, and later led my life into the world of stars.

As I was enchanted by stars, my first wish was to own a telescope which cost a fortune at that time, especially for a poor farm boy. At the time there were only two companies that sold telescopes in Japan. I ordered their catalogues and read them over and over again and eventually had memorised their entire contents.

The more I realised that I could not afford one, the more my longing for a telescope became. Day and night I was thinking of stars and telescopes.

Owing to geographical conditions it rained much where I was born, and I took it for granted. Later, however, after living in different areas, I came to realise that my home village had shown me a tremendous beauty of stars. Shades of the Galaxy, probably around Sagittarius, framed with dark mountain ridges, formed enough view to impress and attract my heart, especially because it did not occur very often.

Not being able to own a telescope, I could only observe shooting stars and the zodiacal light with the naked eye.

A boy in a rural village had come to love the stars. He was alone but gradually became acquainted with many other star lovers, and some of them gave him much influence in later years. These included Dr Kazukiyo Yamamoto of Kyoto University, the chairman of the Astronomy Club. Mr Kaname Nakamura is the astronomer I admire most, but I never met him.

I first observed a meteor shower on April 22, 1932. Under a bright sky lit up by an almost full Moon, I recorded five Lyrids in two hours. I submitted this record to Professor Kohjiro Komaki of the Astronomy Club, and it was printed in the *Journal* of the club. I was so happy I read it over and over again. Since then the Lyrids have become the most friendly meteors of all to me. Afterwards I went more into the observation of meteors and recorded the largest number of anyone in the Club.

Besides meteors, I began observing the zodiacal light without a telescope, instructed by Mr Tsuneo Watanabe. At the same time I came to know Kenji Araki. In later years I went all the way to Mount Ari in Formosa and to Hokkaido and observed the zodiacal light when I went to see solar eclipses.

But still I really wanted a telescope, so finally I decided to buy a 28-mm single lens and a 25-mm Ramsden eyepiece. The former cost 3 yen and the latter 2 yen. They were extremely expensive for me, but it was my first and most memorable telescope purchase of my life.

With these lenses I soon assembled a homemade telescope with which I observed primarily a waning Moon. Even today I still cannot forget the beautiful heart-taking sight of the Moon. It probably was the first and most genuine experience so deeply engraved into my heart and it determined the direction of my life in later years.

About the middle of November in 1930, I was reading a newspaper and found an advertisement of a book entitled 'On Comets' written by Shigeru Kanda of the Tokyo Observatory. For a rural mountain boy, there was no other means to acquire such a book but to send a mail order cheque directly to the publisher. The nearest post office was 8 kilometres from my home, but one day I walked all the way to send a cheque. A note that I wrote on the back of the book cover now tells me that I received the book from the Sanseido Publishing Company on December 25.

Under the dim light I devoted myself to reading the book which was filled with heart-shaking and irresistably charming series of words:

Number of comets discovered:			Total discoveries by country:	
France	Messier	13	France	149
France	Pons	28	USA	84
Germany	Winnecke	12	Germany	81

Italy	Tempel	17	Italy	40
USA	Swift	12	England	20
France	Borrelly	12	South Africa	12
USA	Barnard	21	USSR	11
USA	Brooks	21	Japan	2
USA	Perrine	13		
France	Giacobini	12		

Note that the number of discoveries from Japan is incredibly small. Moreover, these two are both reappearances of periodic comets.

And it continued as follows:

Discovering comets has additional significance. Astronomy is the study of natural phenomena which constantly change, and to record them is quite important for further study in the future. If nobody searches and all the discoveries were made by pure chance, some phenomena might never be recognised. Even if they are found, delay may threaten the advance of astronomy.

Modern astronomy largely depends on young observers who search for, observe, and discover comets, novae, variable stars, and meteor showers. The author hopes that some of the eligible readers who are interested in observation would discover as many comets as possible and dedicate themselves to the development of world astronomy.

How many books does a man read in his life? I have read many; some were beneficial, some were of no use, and many I just flipped through. Whether good or bad, they all influenced me to a certain extent. But to me, no book was so important as 'On Comets'. Other sections that I would like to quote read:

The person who discovered the largest number of comets in the shortest period of time was Perrine who found 13 comets in less than seven years (1895–1902). Especially from November 1896 until March 1898, the discoveries of all six comets were made by him.

Between 1886 and 1890 comet discoveries were concentrated owing to the enthusiastic Barnard. And then Perrine followed him. It is clear that the harder one looks, the more comets are to be discovered. This fact implies the existence of many unknown comets.

A country boy could somehow understand these words and the importance of comet searching, but then what could he do with a 28-mm homemade telescope that magnified 30 times? Yet, even such a crude telescope served well for observing sunspots and the satellites of Jupiter. So it should capture some comets also. At any rate, he had to do with it since it was the only telescope he had.

From December to the beginning of March, it snows much in my home district. In April all the mountains reappear, and by the end of April the cherry blossoms start to bloom in the mountains and the farmers start to get ready for rice planting. And this was the season I started searching for comets with my 28-mm telescope.

As I reflect on it now, what a reckless start it was! I had neither preliminary knowledge nor a star map at that time. A country boy did not even know the existence of a star map. Where should he point his telescope? Where should he search? The fact he knew for sure was that he was facing the celestial dome where uncountable comets have appeared and disappeared since ancient times, and where

Messier and Barnard had struggled. But as for comet searching there is no distinction between newcomers and veterans.

One night I picked up a clear comet-like figure with a tail. I was so excited about my discovery, but it turned out to be a ghost of Venus. I felt quite miserable about my first mistake.

I recall clearly my first discovery that I made without a star map. From 1941 to 1946 I served as a soldier in the battlefield of the Malay Peninsula. What amazed me there were the beautiful stars in the southern Milky Way, the Magellanic Clouds, and especially the Southern Cross which I had never seen from Japan.

One day I found a discarded 80-mm lens and made it up into a telescope of 20 power, and I began searching for comets. My idea was that I might be able to inform my family of my existence by discovering comets, as it was difficult to let anybody know if I were still alive in the battlefield.

The Changy district of Singapore Island where I was serving, surrounded by rubber trees, was too quiet a place to believe that it was in the field of war. I set up my telescope in the front yard of the military base.

Every day after the daily squall, the sky cleared up and gave perfect conditions for observing. My problem was that I did not have a star map. The only way to distinguish comets from nebulae was to observe them several times and look for their motion.

On the night when I started observation, the first nebula-like object of 9th magnitude came into my field of view. In time it disappeared behind the forest of rubber trees. The next day I waited for the night restlessly. Finally it came and I pointed my telescope at the position of the nebulous object, trying to stay calm. When I caught the nebulous object, I found to my surprise that it had moved from the previous position. I was so amazed that it took me a certain time to become fully conscious of the fact, but I finally decided that it had to be a comet.

Later, courtesy of many people, this discovery was reported all the way from Singapore to Tokyo. However, it turned out to be the periodic comet Grigg–Skjellerup and not a new one. It was reckless to search for comets without a star map, and it still makes me wonder how I could have found a comet in such a short time.

In June 1946 I returned to Japan and started observing again.

In 1949 I was using a 15-cm telescope mounted on a hill behind my lodging house. One night I was walking up the hill through a pine grove using a small flashlight. Suddenly I was startled by a huge bird that took off in the pitch darkness with a noisy flutter. I reached my observation hut with my heart still beating fast and opened the roof. Just then I looked up at Perseus and saw Comet 1949l shining clearly with a long tail at 4th magnitude. It was the only comet I have ever seen with my naked eye. I always remember the splendid Honda–Bernasconi comet (1948g) and the sound of heavy flutter of the bird together.

In the same way, on September 19, 1965, when the Ikeya–Seki comet was discovered, I was also looking at the same sky cleared by a typhoon. To my pity, the

Fig. 5.8. Minoru Honda and some of his disciples. From left to right: unidentified, Tsuruhiko Kiuchi (co-discoverer of Comets 1990b and 1990i), Mr Honda, and Sigeru Furuyama (discoverer of Comet 1987f$_1$).

very position of the comet was behind a bush in the neighbour's yard. I still remember the dark shadow of the bush.

In December 1934 a bright nova appeared in Hercules. Every night it shone attractively near Vega in Lyra, and gradually captured my heart when I was about to step into the celestial world. I believe that was the start of my serious interest in novae.

In 1936 a nova appeared in Lacerta in June and another in Sagittarius in October. They both excited me very much, especially because they were discovered by Japanese. From then on I began hoping to discover a nova myself. In 1967 a nova appeared in Delphinus, and also the one in Cygnus in 1975 was magnificent.

At any rate, every clear night as I open the roof of my observatory hut, the stars are always waiting there graciously. I do not have to struggle at all, but they themselves are just waiting and inviting. I can never think of a better welcome. To such kind stars, why do I have to dare observe or search?

I just want them to teach me something. To enquire the stars, comets, and novae in the vast universe still seems to be somehow arrogant, yet the stars, one by one, kindly visit my hut again tonight . . . for me who is waiting here alone. [Figure 5.8.]

[This contribution, written just a few weeks before Minoru Honda died, was kindly translated by Toru Hayashi. Honda's words were written specifically for the readers of this *Guide*, but following a request made to Dr Jun Jugaku of Tokai University, they were printed (in the original Japanese) in the *Astronomical Herald*, the journal of the Astronomical Society of Japan.]

UNTITLED REMARKS
Albert Jones, OBE, Nelson, New Zealand

Although I did a little astro-photography in the late 1940s, I have done none since, but prefer to work visually – since 1948 with a 12.5-inch reflector. The f/5 mirror was made by F. J. Hargreaves (England) – the eyepieces of the telescope and finders are from war assets anti-aircraft predictor telescopes – the 78-mm OG [objective glass] for one finder was brought back from World War II by a cousin (I believe it came out of a German Tiger tank) – the 45-mm OG from Barr & Stroud rangefinders. The square tube was made from aluminium alloy angles covered with thick builders' wallboard. Don't laugh – it works – the first tube I made was covered with thinner dense wallboard and the mirror cooled and dewed over twice in a few years, but in nearly 40 years the insulating wallboard has kept the mirror from dewing over, even though the humidity in Nelson [on the north coast of the South Island] is higher than in Timaru [on the east shore of the South Island]. As the telescope is used for VSO [variable star observing] and the occasional comet, I have no need for a drive, so push the thing about by hand – likewise as I locate star fields by star hopping, there are no setting circles. The fork mounting could be used as an equatorial, but to obviate the awkward angles that a Newtonian eyepiece can be at away from the 'meridian', I turn the whole thing (on wheels) so that the 'north' end of the polar axis points in the direction I wish to look; then there is the eyepiece nice and handy – unorthodox, but again it works. [See Figure 5.9.]

However, when I located Comet 1946 VI I was using a 5-inch refractor on a tall tripod. That morning I had done some VSO, then swept for comets in the southeast, stopping about daybreak to have a look at the U Puppis field which by then had risen in the east. I should have grabbed a box to reach the finder, but to save time I roughly pointed the telescope and, realising it was too high, started sweeping downward to find the U Pup field. After a few sweeps I came across a fuzzy thing which was not marked on *Norton's Atlas* (the best Atlas I had then). I made a quick field drawing and connected it to Norton stars but dawn washed out the view before motion was established. Wanting to make sure, I waited until next morning but clouds came over before the comet rose – I waited impatiently for the sky to clear and luckily clouds started to roll away just before daybreak. First the discovery position showed, and nothing was there so it was not some permanent scenery, and shortly afterwards (though it seemed ages then) I was able to see the comet again, get another position, then call Carter Observatory.

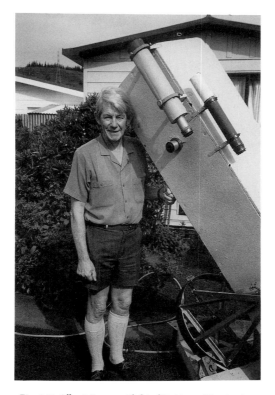

Fig. 5.9. Albert Jones with his f/5 32-cm Newtonian.

Independent discoveries of comets P/Kopff and P/Grigg–Skjellerup were made by looking at positions from BAA *Handbooks*.

SN 1987A

In one of Frank Bateson's chart series, a chart by Mati Morel showed three stars near the Tarantula Nebula that might be unstable sometimes: NSV 2499, CPD − 56° 420, and HDE 269858. I had looked at them fairly often for about a year. One night I had been observing T Cha when I noticed clouds coming over from the west, so I poked the telescope to the LMC to 'do' those 3 stars, but when I looked through the finder there was that bright bluish thing – yes, I was looking at the right place, the Tarantula Neb was close by. I hurriedly marked the position on the chart, but before I had time to sort out comparison star and make an estimate, clouds beat me to it – that was Feb. 24d 09h 13m UT. I waited (again impatiently) for the clouds to roll away, so that I could give a mag. estimate to Frank [Bateson, Director, Variable Star Section, Royal

Astronomical Society of New Zealand], but when it seemed the clouds were going to stay, I phoned him, gave him the position, and said it was about mag. 7 (I underestimated it). When I came away from the phone, there were a few stars showing, so I had another impatient wait for the clouds to go away, and at 10h 23m I wrote down '?LMC' at 4.8 mag. Then I phoned back to Frank, who then phoned the AAT at Siding Spring as he could not raise the IAU Bureau, and the Post Office in NZ had stopped taking Telex messages at night. Not knowing if anyone else was watching the 'thing' I kept on observing it, then a few variables, then the thing, then more variables until 13h when I had a short snooze before resuming observing at 14h 43m until dawn when I observed Comet Wilson – it had been quite a night. Frank phoned at breakfast time with the news that it had been seen in Chile a few hours before me.

With regard to the OBE awarded in the Queen's Birthday Honours 1987, I was pleased to hear some time after the award that the nomination (by the RASNZ) had to be in *early* February, so that was before the fuss over the LMC thing – I would rather be remembered for 'work' done over the years than for a chance look at the LMC at the right time . . . The OBE stands for the Order of the British Empire and not as someone said 'other bugger's efforts', for I can honestly say it was for my own efforts.

DISCOVERIES FROM THE BACKYARD
Daniel H. Kaiser, Columbus, Indiana, USA

In 1985 I read Ben Mayer's book *STARWATCH*, in which he described his PROBLICOM method of conducting a photographic sky patrol. This entailed repeatedly photographing areas of the night-time sky and comparing the photographs in search of novae. I quickly realised this was a project well suited for me as I already had all the necessary equipment, except for the comparator.

My observing site is my backyard which suffers from light pollution. Not letting this deter me, I was soon taking photographs with my 35-mm camera piggy-backed on an 8-inch Celestron telescope. [See Figure 5.10.] Experimenting with different lenses and exposures, I soon settled on my 135-mm lens and 5-minute exposures. This covers about 10 by 15 degrees to 10.5 mag. when photographing overhead.

Even though my light-polluted horizons restricted the areas I was able to photograph, I repeatedly photographed whatever was accessible. It wasn't long before I was finding lots of things: airplanes, satellites, asteroids, and variable stars. Lots of variable stars.

After several months and as many rolls of film, I had learned how to identify everything I was finding. Variable stars could be recognised as such by looking at earlier photographs to see if a star was recorded at that location. Airplanes and satellites would not appear on both slides taken on the same night (I quickly learned always to take two shots of every field). Asteroids will move in a couple of days, and so will identify themselves by their motion.

Fig. 5.10. Dan Kaiser with his variable star camera piggy-backed on a 20-cm telescope.

I was finding lots of things. However, I was not finding any novae.

Not being the kind of person to give up easily, I continued to take my photographs and decided to look harder at what I was finding, which were lots of variable stars. With the aid of the *AAVSO Variable Star Atlas*, I was able to identify most variables, not only as variable stars but also with their given designation.

Then one day, a little over a year into my project, I found three variables not on the *AAVSO Atlas*. Three in one day. I contacted Marvin Baldwin of the AAVSO, and we searched his copies of the official variable star lists. My three stars were not listed in any catalogue. Could it be that these were unknown variables? Marv's reaction was . . . 'Well, it *could* be'.

This served to whet my appetite. I knew it would only be a matter of time before I found my first nova. Continuing to photograph my star fields, paying extra attention to my three 'could be' new variables, weeks turned to months. No novae.

In March of 1988, my big discovery happened. However, it was not what I had been searching for. It was not a nova. While searching a field in Gemini, I noted an 8th magnitude star had faded considerably since my last photograph. A quick look in the *AAVSO Atlas* showed no variable star at this location. Nor was it located in the *General Catalogue of Variable Stars* (GCVS), Marv's variable star catalogue which he had most generously loaned me. I called Marv.

Two days after the discovery photo was taken we were both surprised to find that the star was not only still fainter than normal, but it had faded even more. There was no denying it, this was not just a 'could be' discovery. Something was happening that had never been witnessed before. We watched and recorded its rise back to normal brightness. It took a full week for the star to recover from what Marvin had immediately recognised as an eclipse. This star was an unknown eclipsing binary with a very large amplitude, 1.8 magnitudes. And a very long eclipse, 12–14 days.

After a study of Harvard College Observatory's photographic plate collection revealed a period between eclipses of 1258.56 days, 3.45 years, the IAU assigned the official designation OW Geminorum.

For me this is better than a nova. Why? In September 1991 I can watch it happen all over again!

As of this writing, June 1990, I have shot just over 100 rolls of film and spent I don't know how many enjoyable nights bathing in and recording starlight. I have found 15 previously unknown variable stars, but still no novae.

Would I recommend this pursuit to anyone? You bet I would!

The PROBLICOM really is an easily constructed device, and the old saying 'If I can do it, anyone can!' does apply here. Besides the two projectors, all that is needed are some plywood, hinges, turnbuckles, and a few screws.

If anyone builds a PROBLICOM, and photographs some star fields *systematically*, their problem is not going to be finding blinking objects. Rather, their problem will be identifying the blinking objects they are bound to find.

After 4 years of pursuing the elusive nova unsuccessfully, and serendipitously discovering several new variables along the way, my heart still beats a little faster every time I see an image blinking back at me. It does not matter if it turns out to be the brightest asteroid known to man, the eighth major planet, or a known Mira-type variable. It is still fun to find and identify these transients among the stars.

The camera and mount

Any 35-mm camera that has a shutter that can be locked open for time exposures will work. Cameras with a mechanical shutter are preferred over auto cameras. The mechanical shutter will not fail when the battery is weak.

Almost any lens will work, although lenses in the 85 mm to 135 mm range are recommended. Shorter focal lengths tend to record the brighter stars only. Longer focal lengths restrict the field coverage.

As for the mount, almost any equatorial mount with drive will do. Since patrol photography is basically piggy-back photography, the tracking accuracy of the mount is not highly critical. As long as the mount is reasonably aligned and driven, it will suffice for our purposes. I use a 135-mm lens and have never had to guide my exposures.

The film

Ask any astronomer which film is best for astro-photography, and you will undoubtedly hear hypersensitised Kodak Tech Pan 2415, and they will be right: It will record finer detail and fainter stars faster than most colour film. So, you may ask, why don't I use it?

Well, if I go to the local 'K Mart' and ask for Kodak Tech Pan 2415, preferably hypersensitised, I get blank stares.

When I started out to hunt for novae I was already using Kodak Ektachrome 400 colour slide film for astro-photography. I like it because it is readily available and I can develop it myself without a darkroom using Kodak's 'Hobby Pac' developing kit. No special equipment is necessary, i.e., hypersensitisation equipment.

Ekta 400 is rather fast and also relatively fine grained. An advantage of colour film over black and white is just that; it shows colour. If the blinking object is white, then it is probably an eclipsing binary or Cepheid. If very red, then most likely it's a Mira or semi-regular variable. Orange could be a nova.

Now that I have a collection of over 2000 slides to compare new photographs to, I find it hard to consider changing.

Hints

These hints are suggestions that I wish I had access to when I first started my patrol programme. Most are hard-learned lessons. Some I leaned from my mentor, Marvin E. Baldwin, AAVSO Eclipsing Binary Chairman, to whom I will be forever grateful.

Marvin introduced me to the GCVS and the NSV (New Suspected Variables). Also the practice of triple checking *everything*.

1. Always, ALWAYS take two photographs of each field. Sequentially, not simultaneously. You will never regret the use of the extra film. It will help exclude film flaws, flashing satellites, airplanes, and the like. It not only gives you confirmation photos but will save you the embarrassment I endured when I met Dr Martin Burkhead, of Indiana University, on a Sunday afternoon to show him my first discovery (a film flaw).

2. Take accurate notes. Record exposure number (roll/frame), target, double date (e.g. Jan. 5/6 1990), time (UT or local), lens, f-number, exposure duration, and film type. I have made up a form which I use. This assures that I record all the necessary data even if it is 2 a.m.

3. Test your lenses for proper focus. Shoot a series of exposures changing the focus ever so slightly each time. Note the position of the focus indicator for each shot. I found that the infinity focus on my 135-mm lens is not at the infinity mark! A sharply focussed frame will increase your limiting magnitude.

4. To identify what you have found blinking on your PROBLICOM, you will need several catalogues, or at least access to them. I use the *Smithsonian Astrophysical Observatory* (SAO) *Star Catalog*, the *General Catalogue of Variable Stars*, 4th edition (GCVS), and the *New Catalog of Suspected Variable Stars* (NSV).

   The *SAO Star Catalog* contains over 250 000 stars down to about 9 to 9.5 magnitude, with accurate positions and pretty accurate magnitudes in the blue and/or visual. The GCVS is the official list of known variable stars and is indispensable if you hope to identify anything new. The NSV is the latest (1982) list of suspect variables.

   If you are lucky enough to have access to a computer, all of these catalogues are available in an IBM-compatible program called SUPERSTAR, available from picoSCIENCE, 41512 Chadbourne Drive, Fremont CA, 94539. The program costs less than the total of the catalogues it contains. It includes an asteroid catalogue as well. It will certainly speed the identification of blinking objects considerably.

5. A good star atlas is also very handy. I recommend the *AAVSO Variable Star Atlas* because it is based on the SAO and already has the location of most variable stars brighter than 10th magnitude marked.

I have never used black and white film for my search efforts; I chose Ektachrome 400 for the following reasons . . .

1. I was already using Ektachrome 400 for my aesthetic (read that pathetic) astrophotography.

2. Having no darkroom, and not being willing and/or able to invest in one, I can develop E400 with the 'Hobby Pac'.

3. I sometimes mix patrol photos and 'pretty' astro-photos on the same roll. [But see the remarks at the end of the contribution.]

I must tell you that many of my patrol photography decisions were made for economic reasons rather than scientific ones, i.e., the above-mentioned darkroom, the use of a 135-mm lens (it's what I had), shooting from a light-polluted backyard, not using hypersensitised film, etc. If you do choose to follow my suggestions, you may want to call it 'Astronomical Discovery on a Shoe String' or 'Doing It on the Cheap'.

[He may come to hate us forever, but Mike Collins and I convinced Dan to start using hypersensitised Technical Pan film. In his most recent letter (March 1991) Dan reports that he is a happy man.]

UNTITLED REMARKS
David H. Levy, Tucson, Arizona, USA

Although my interest in comets dates back to the sixth grade, I did not discover how rewarding comet hunting could be until I read Leslie Peltier's marvellous book *Starlight Nights*. Much more than an autobiography, this book accurately portrays the spirit of what amateur astronomy is all about. If I were to be marooned on a desert island but with one allowed item, I would want (besides a telescope, computer, and modem for reporting comets) a copy of *Starlight Nights*.

Comet hunting began for me on December 17, 1965, when I spent 10 minutes hunting between Castor and Pollux. During the following few years I learned a lot about what to do and what not to do when comet hunting. The first mistake was to spend too much time hunting under a city sky; although one could see stars down to 11th magnitude, the low surface brightness of most comets would preclude my finding anything fainter than 7th or so. Despite this, in September 1968, while hunting from the middle of Montreal on a night near full Moon, I did pick up Comet Honda 1968c. Although the comet was long since discovered, I had no idea where it was, and the shock of actually picking up a comet while comet hunting was a real encouragement.

Although my first years of hunting were done with the relatively light-polluted sky over Montreal, there were several glorious summers of observing from the Adirondacks south of the city under pristine skies. By 1979 I had decided to relocate to a place where a dark sky would be available to me year round, and in the last months of that year I enjoyed a period of almost 60 uninterrupted clear nights. With the dawn of the 1980s and after 15 years of comet hunting, I was finally ready to find one!

On November 13, 1984, I cut short a wonderful dinner date to rush home to a sky clearing after several nights of clouds. 'O.K. stand me up,' my friend Lonny Baker had said, 'but you'd better find a comet for me tonight!' One hour and seven minutes of comet hunting later, I found the strange sight of Aquila's open cluster, NGC 6009, with a fuzzy patch next to it. The contrast between cluster and fuzz was so beautiful I knew that something was out of place, for such beauty would have appeared in all the astronomical picture books. Within ten minutes I had my answer, for the fuzzy patch had moved! After reporting the comet I called Lonny. 'Well, did you find a comet for me tonight?' she asked. When I answered 'Yes,' she laughed, and when I told her the magnitude, she laughed again. It was only after I provided the position and direction and rate of motion that she stopped laughing: 'My God, you're serious!'

In January 1987 I was beginning a new year with two resolutions, one to finish my almost-completed book on observing variable stars, the other to discover a comet. On the morning of January 5 I finally completed the book, and set the computer to prepare the manuscript for printing. With an hour to go till dawn I checked the sky, saw that a small break in the clouds had appeared, and began a brief comet hunt. With both dawn and more clouds rapidly approaching I hunted in the southeast sky until the sudden increase in humidity told me that I'd better stop soon. I did, but not before moving one more telescope field. Near the edge was a diffuse fuzzy patch. I quickly made a sketch before a cloud covered it; I closed the telescope and observatory and rushed inside; ten minutes later it was raining.

It was two days before I could see the object again, although during the interim I had checked the *Palomar Observatory Sky Survey* and found nothing in its position. Now it was a bit more than a degree to the south; two hours later an IAU *Circular* announced the discovery of Comet Levy 1987a.

Each of the other finds has its own special story. For example, Comet 1989r found its way to the Earth on the night of the Voyager 2 Neptune encounter, and I always enjoy remembering that unusual observing session whose time was divided between going inside to watch new pictures of Neptune and Triton, 'live' in my living room, and then going outside to catch a new comet.

The sixth comet came only a few months later. It began near the end of a long observing run with Steve Larson at the 61-inch telescope on Mt Bigelow. We had been observing Comet Austin, but since there was not as much to observe as we had expected, Steve suggested I stay home Saturday night and find another comet; 'We need another one to observe!' The next night, Saturday May 19/20 began with a pleasant evening conversation I was having with Walter Scott Houston. As the evening ended I mentioned that I would be comet hunting until moonrise the following morning. 'If you find Number Six,' Houston said, 'you will be tied with Mellish,' (the mid-western American comet discoverer and optical craftsman). With a clearing sky, I began hunting in the southeast in a leisurely, up-and-down pattern until the brightening horizon indicated imminent moonrise. Should I stop at this point? By now the sky was extremely clear despite the Moon, so I decided to move the telescope to the north, avoid the Moon, and continue hunting. Two or three fields past Alpha Andromedae I noticed an obvious fuzzy patch; I knew this part of

Fig. 5.11. David Levy at right; Rod Austin at left; Frank Lopez, centre.

the sky thoroughly and instantly a 'red alert' went off in my brain. A second look revealed a tail!

The real surprise came when I observed the new object the following night and found that it had moved only an eighth of a degree, several times slower than the $\frac{1}{2}-1$ degree daily motion which is typical for most comets. With the publication of the first orbit for Comet Levy 1990c, I knew the reason why; this comet was out in the asteroid main belt and approaching the Earth fully five months before perihelion, and heralding an interesting display in the August sky.

Several different aspects of comet study interest me. My Master's degree thesis in English literature had to do with the poet Gerard Manley Hopkins' interest in comets and specifically his observation of the Comet Tempel–Respighi 1864 II and his including it in one of his poems. From 1985 to 1989 my work with the International Halley Watch's Near Nucleus Studies Net improved my understanding of the professional research now being done on comets, and in CCD observations of comets. Observing with Carolyn and Eugene Shoemaker on the 18-inch Schmidt camera at Palomar, plus working with my own 8-inch Schmidt, have helped me enjoy the photographic aspects of comet work. Finally, as part of the work I have done for a biography of Clyde Tombaugh, the discoverer of Pluto, I found that he had also discovered a comet on 1931 plates. Since these plates were a year old at the time, Tombaugh had never reported the comet, and I spent much time and travel searching old plate archives for other images. Although I never found any, I enjoyed travelling back in time to an era where comets got caught on glass plates, their frozen photons waiting for a future observer to unlock them.

[Figure 5.11 shows David with Rod Austin and amateur Frank Lopez. David has recently begun to participate in the Shoemakers' Palomar programme. See Carolyn Shoemaker's contribution later in this chapter.]

HOW CAN I AVOID MISSING OUT ON FINDING COMETS?
Don Machholz, San Jose, California, USA

In 1977 I studied this problem: 'Why was I missing out on discovering comets even though I was probably the most active comet hunter in the world (500 hours/year)?' Occasionally a comet would be found, but by someone else first. From this study I determined that three factors will cause you to miss comets.

1. Magnitude: the comet must be bright enough to be seen by you. If you are missing them for this reason, work on improving your eyes, sky, or telescope (see below), sweep slower, or concentrate more.
2. Position: the comet must be in a part of the sky you are sweeping. If you are missing comets because of this, cover more sky and spend more time sweeping.
3. Time: the comet must be seen by you before it is discovered by others. If you are missing comets due to this, get out more often and especially after the Moon clears from that part of the sky.

Of the three above factors, the most important seems to be magnitude.

How can I see better

Sometimes people ask this question, or, 'What telescope should I use to search for comets'. I usually suggest they use what they have and are familiar with, then get something else if they want to continue hunting for any length of time. But these three factors affect how well you see faint objects:

1. Eyes: experienced, dark-adapted eyes see better.
2. Skies: you need dark skies.
3. Telescope: you need something giving you good contrast, light grasp, portability, and stability.

Binoculars for comet hunting

I do not often have the money to buy commercial equipment for comet hunting, and even when I do, I decided long ago that I would not make my astronomy a financial burden. So I end up building everything myself. And I'm not real good at building things; usually I use wood (easiest to work with) and pipe mountings. May I point out that my whole binocular system, used to discover comets 1986e and 1988j, was built for under $400; this is less than the cost of even one Jaeger lens. Here is their description.

The lenses (actually aerial telephoto lenses) are 127 mm across, the binoculars magnify 27 times, and they weigh over 100 pounds [45 kg]. In April 1983 I had seen these lenses being sold at our astronomy club auction. At first sight I wasn't overly impressed with the lenses; in fact, I left the auction before it ended and never saw them bidded upon. In the following week I began to consider making a pair of large binoculars with them. I called their owner, Steve Greenberg, and he said I could have them for $50 each, a bargain price. I spent the next two weeks planning and designing a binocular system using secondary mirrors. When I was convinced that it would work and be completed for under $300, I bought the lenses and ordered the mirrors.

Following two weeks of construction, I had a pair of large binoculars. Everything is enclosed in a plywood box measuring 92 cm long, 56 cm wide, and 30 cm high. Each of the two light paths has two elliptical mirrors, one measuring 66 mm in minor axis, and the other measuring 47 mm. The light strikes the first mirror, right-angles toward the centre of the optical system, then strikes the second mirror, exiting the eyepieces through the back. The eyepieces are surplus 30-mm Plössls that I already had and provide a 3.4 degree field. They are set into plastic PVC pipes making focussing difficult. Originally I was able to vary the distance between the eyepieces, but then I changed them to a fixed position so the alignment would remain more stable. [See Figure 5.12.]

The contrast is good in this instrument. I can see 'star clouds in every constel-lation'. However, I use them for about one-third of my comet hunting. For most of my searching I use my 25-cm, f/3.8 reflector at 32 power. [See Figure 5.13.] The field of view has been stopped down to 1.6 degree square.

The binoculars and the telescope are roughly equal at picking up faint diffuse objects. Although the telescope is larger and gathers more light, using two eyes in the binoculars helps to acquire faint objects too. I can cover the skies faster with the binoculars, but it is not possible to boost magnification to check out suspicious objects. Each instrument has its advantages and disadvantages.

On May 12, 1986, I began comet hunting session No. 1471 at 1:50 a.m. by scanning, or sweeping, the sky high in the east. While looking through the binoculars, I would slowly rotate the binoculars from right to left, looking among the pinpoint stars for small fuzzy clouds. Most of the time these 'small clouds' turn out to be clusters of stars, galaxies, or nebulae. But one could be a comet. At the end of each sweep I swing the binoculars back to the original point, lower them one to two degrees, then repeat the process. By starting high in the eastern sky and slowly working my way downward, I should reach the horizon in about two and a half hours.

Fig. 5.12. Don Machholz's homemade 27 × 127 binoculars.

Time went by very quickly, as I was quite busy 'working' the northern sky. The Milky Way looked fine during the early part of my sweeping, with several star clusters and nebulae catching my attention. My eventual downward motion freed me of the Milky Way into rather bare sky with few stars. Here I picked up a little more speed and by 3:45 a.m. I swept up the Great Andromeda Galaxy, M31.

Worldwide, perhaps no more than a dozen comet hunters are sweeping the sky on any particular morning. Because of the time and discipline needed, few amateur astronomers take up this hobby. I began on Jan. 1, 1975, with the idea of trying to find a comet. Besides, having some ten years' experience observing the Moon,

Fig. 5.13. Don Machholz at the focus of his 25-cm reflector.

planets, asteroids, galaxies, and clusters, then photographing them, I wanted a programme that would challenge me and encourage me to look through the telescope. Monitoring variable stars, chasing asteroids, or hunting comets would all fulfil these requirements. Noticing that few Americans hunted comets, but that lots of second-hand advice was written on the subject, I decided to hunt comets and see for myself what it was all about. Besides, I had read that it took only about 300 hours of searching to find a comet. I felt I could benefit from 300 hours of looking through a telescope even if I didn't find a comet.

After 1700 hours of searching, I discovered my first comet (Comet Machholz 1978l), and 1742 hours later I found my second (Comet Machholz 1985e).

On the morning of May 12, 1986, it did not occur to me that it was exactly fifty weeks since I had found Comet 1985e, but it was. Half-way through a sweep, something near the top of the field caught my attention. I stopped sweeping and examined it. It looked like a small fuzzy object, but perhaps it was a small group of stars. I placed it in the centre of the field of view. It looked like a diffuse object, round, and just within the limits of visibility. It was 3:52 a.m. The radio was playing 'Against All Odds'.

Checking its position I determined that it was two degrees south of the Andromeda Galaxy. There were no galaxies or clusters listed on the maps; now I had to check to see if it was moving.

I drew a map of the region, placing an 'x' at the comet's location. I then resumed comet hunting. At 4:17 I returned to the object's location and compared it with my map. It appeared to be moving, but I couldn't really be sure. I again resumed sweeping. At 4:39 the sky was beginning to brighten as dawn was approaching. I re-found the object and was very pleased to detect motion. It was a comet!

Later, from home, I phoned the Smithsonian Astrophysical Observatory in Cambridge, MA, and reported the comet to Daniel Green. He had not heard of any other objects in the area and together we felt this was a 'reasonable suspect' for a comet. [It turned out to be Comet Machholz 1986e.]

This, my third comet, turned out to have an elliptical orbit, and thus the comet 1986e will return every 5.4 years. This proved to be quite exciting for professionals as periodic comets with this short an orbital period are rare and of all the short-period comets known, this one comes closest to the Sun — 19 million kilometres. At its farthest point it goes just beyond Jupiter's orbit. The path of this comet is tilted 60 degrees to the Earth's path around the Sun; it enters the inner solar system from the south and exits in the north.

[Note: Don Machholz writes a regular column for the ALPO *Recorder* and circulates a newsletter 'Comet Comments'. Some of the above was reprinted — with his permission — from this series. In these writings he has described his own comet discoveries as well as those of others. Additionally, he frequently writes on historical subjects and on the statistics of comet hunting. These articles are highly recommended for those who wish to make a serious business of comet searching. See Appendices H and I.]

DISCOVERING MINOR PLANETS
Brian G. W. Manning, Kidderminster, Worcs, England

The possibility of searching for minor planets was suggested to me by Robert McNaught [see next contribution] who pointed out that Japanese and Italian amateurs had already been successful in this field. At first I must admit that I was doubtful about the possibilities of doing this in England due to our light-polluted skies and long periods of cloud in winter. The latter might make a series of observations difficult in the event of a discovery, a single observation being of little use, of course. In the summer the problem is the low altitude of the ecliptic at 52 degrees north latitude. However, in the spring of 1989 I successfully photographed my own asteroid (3698) Manning and also (1657) Waterfield. On the Waterfield negative there were three more 16th magnitude asteroids, but all already numbered, which made me think that Rob McNaught was probably correct, and that it would be worth attempting to discover a new one.

In the autumn of 1989 I noticed that Comet P/Schwassmann–Wachmann was near the ecliptic and travelling at near-asteroid rate; therefore, if I photographed it, there was a chance that I might get a minor planet, and if not, I at least would have the comet.

On the 4th of October I made two 25-minute exposures with my 25-cm, 1.9-m focus telescope [Figure 5.14], and to my delight recorded four asteroids as well as the comet. The next night, follow-up exposures were made, but due to bad aiming, one asteroid was out of the field. The others, however, were well-recorded.

The positions were measured on my homemade measuring machine [Figure 5.15], and sent to Dr Brian Marsden at the Minor Planet Center over the electronic mail system by Guy Hurst, editor of 'The Astronomer'. The reply was that none of the positions corresponded with asteroids with permanent numbers or observed at more than one position. This caused considerable excitement as may be imagined. The asteroids were given the designations 1989 TE, 1989 TF, and 1989 TN$_1$, and eventually the single-night observation of 1989 TD$_{11}$. (Gareth Williams explains that the numeral after the two letters indicates how many times the alphabet has been cycled through in the relevant half-month period. So 1989 TD$_{11}$ would be the 279th discovery of the first half of October 1989.)

1989 TF was identified with 1968 OF, and because a previous observer had a better series of observations than mine, this observer will become the official discoverer when the asteroid is numbered. 1989 TE was identified with 1982 TB, but when it has been observed at future oppositions, I shall be listed as the discoverer. 1989 TN$_1$ may also be eventually credited to me. The fourth one is just an observation in the MPC computer waiting for someone to find it again.

There was considerable interest in these discoveries because it was almost exactly 80 years since the last minor planet discovery in England. There were, however, two single-night discoveries at St Andrews University, Scotland, by F. P. J. Smith in 1986; these were 1986 XP$_5$ and 1986 XQ$_5$.

In order to be named as the discoverer when an asteroid is numbered, one must have observed it sufficiently well at one opposition to make it possible for a good orbit to be computed so that previous observations can be traced. Before an asteroid can be numbered and named, it must have been observed at four oppositions (three if it has been very well observed) in order that a highly reliable orbit can be computed.

As of this writing I now have two numbered discoveries: The first is No. 4506 (1990 FJ) which I named 'Hendrie'; I had five observations from the 15th to the 26th of March 1990, with a sixth observation to clinch the matter on April 28. The second, No. 4751 (1991 BG), named 'Alicemanning' for my wife, was found on the 17th of January 1991 with only one other observation obtained, on January 21, owing to cloudy weather. Fortunately, Rob McNaught very obligingly took follow-up photographs in Australia, and these combined with observations at Tautenberg [Karl Schwarzschild Observatory in Germany] enabled an orbit to be obtained. The Minor Planet Center traced it back to 1954. I was rather surprised to be credited with this one, but apparently the fact that the two observations had allowed other observers to follow it at the current opposition was sufficient.

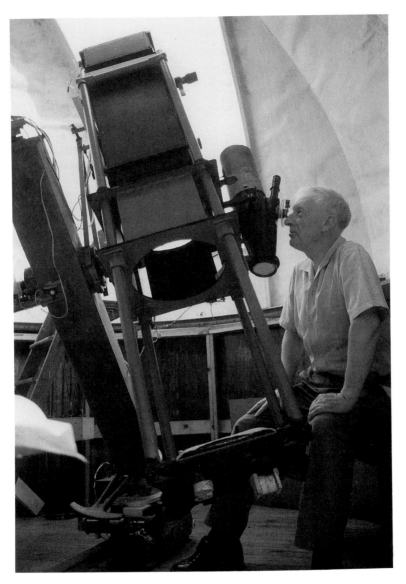

Fig. 5.14. Brian Manning and his 26-cm telescope.

Fig. 5.15. Brian Manning's measuring machine.

Three other asteroids were found in the short spell of clear weather in January 1991. One of them, 1991 AF_1, is rather interesting because Gareth Williams of the Minor Planet Center discovered an observation as far back as 1934 at Heidelberg by Karl Reinmuth, i.e., only two years after the death of Max Wolf, the pioneer of asteroid discoveries by photography. The next observations were in 1958 and 1980, and then a whole lot of observations close to mine in 1991. I was rather pleased to learn that I shall eventually be the official discoverer of this one.

The professional way of finding asteroids is, of course, by detecting the trails they leave on photographs taken with large, fast Schmidt cameras, or more recently with giant CCDs. They can also be detected by examining pairs of photographs with blink or stereo microscopes. Discovery of a new minor planet brighter than 15th magnitude is rare, and mine have been in the range 15.5–17.5 visual mag. For the amateur with light-polluted skies, this raises problems. Exposures long enough to show detectable trails with a fast camera will produce sky-blackened negatives, and long exposures with an ordinary slow reflecting telescope, like my own, will not record trails of faint asteroids. The solution is to track the telescope at the average rate (about 30–35 arcsecs per hour) for an asteroid near the ecliptic and near opposition. Asteroids are then recorded as star-like images, and there is a gain of 2 or 3 mags over trailed images. My guide telescope is an f/5 folded refractor fitted with a stepping micrometre controlled by a home computer. It is essential to make two exposures on the same night, partly as a check on film flaws, i.e., to distinguish the spots from the dots, and partly because a direction and motion can be established to follow up any objects detected. I examine my negatives with a binocular microscope of 8.75 magnification.

Having said that, I note that other amateur asteroid hunters seem to use fast cameras, in the range f/2.5 to f/4. [Brian's telescope is f/7.3.]

Over the period October 1989 to April 1991 I have detected 24 minor planets on my negatives; of these, five were already numbered. In addition to my two numbered asteroids, there are four for which I shall be listed as the discoverer when they have been observed sufficiently to have received numbers. Another five may be credited to me after they have been observed at future oppositions. This is a rather frustrating feature of asteroid discovery: one almost invariably finds them at a perihelic opposition with the result that unless their orbits have a small eccentricity, they are likely to be one or more magnitudes fainter at their next opposition – and most likely undetectable by the amateur. At their next bright oppositon they may have deviated a long way from the first preliminary orbit and be difficult to find.

I use 50 mm by 63 mm pieces of Kodak TP 4415 film 'hypered' in pure hydrogen for 2.5 hours at 50 °C. I DO NOT use cylinders of compressed hydrogen at high pressure because of the danger of a sudden leakage and the risk of fire or a serious explosion. Instead I generate hydrogen electrolytically in one-litre batches as required, and even then, take care that static electricity cannot cause ignition when releasing the gas from the hypering tank.

At the true prime focus of my telescope allowing for masking, I have a field of 1.7 by 1.2 degrees. Coma is not too serious at f/7.3. There is a cover glass to keep damp air from the emulsion.

The measuring machine is very similar to an engineer's toolmaker's microscope. I was fortunate in having carefully preserved a pair of 'inch' screws and nuts with their micrometer heads which became redundant when the laboratory where I worked went metric. Linear ball slides were purchased and fitted to machined iron castings to make the traversing tables for X and Y coordinates. The area covered is 3 by 2 inches (76 by 50 mm). Eventually, I constructed a small circular dividing machine incorporating a stepper motor controlled by my computer, and made some 500 division encoder disks which I fitted to the measuring machine screws. I now have digital read-out and can also transfer the reading to a computer by pressing a switch. Measurement takes up quite a lot of the time spent in minor-planet hunting.

I have not ventured into the field of hunting minor planets with CCDs and will probably not do so. They do, however, solve the problem of the measuring machine which would be a costly item if purchased. Also their at present small field could no doubt be offset if objects fainter than 18th magnitude can be detected.

I should like to express my thanks to Dr Brian Marsden and Gareth Williams of the Minor Planet Center with whom I have communicated via the Starlink Node at the University of Birmingham, England, for which I am also very grateful.

MISCELLANEOUS REMARKS

Robert H. McNaught, Coonabarabran, NSW, Australia

My nova patrol

[Excerpted from a letter written in March 1987.] I cover the whole MW [Milky Way] with a 24-mm lens. This covers ± 30° galactic latitude with potential discovery of mag. 8, possibly mag. 9 objects. I feel the discovery rate for mag. 7 objects present on the photo must be near 100 per cent. In addition, the same longitudes of the MW are covered with an 85-mm lens to ± 10° to mag. 12. I feel discovery is certain to mag. 9, with some mag. 11 variables being found. As you will be aware, some will necessarily be missed due to involvement with nearby stars of similar or brighter magnitude. Note that only about 10 per cent of my 85-mm shots are searched within a week or so, but with the Mag[ellanic] Clouds and the wide-angle shots about 95 per cent are searched, and coverage is about once every two days, with around 4 days a month lost due to moonlight. However, photos are still taken through the full Moon, just in case! These are not searched, only used for reference.

Automatic photography

[Excerpted from a letter written in August 1986.] I use a Canon T-70 with a command back and a Super Polaris [Celestron] mount. As the mount can be controlled via a Skysensor, one can program a list of field centres:

	Camera	Mount
Field 1	Open	
	Shut	
	Wait	Slew
Field 2	Open	
	Shut	
	Wait	Slew

The camera and the mount must be synchronised and, of course, backup photos can be taken. For these I usually use a 2-second wind-on, but this can be arbitrarily long with the T-70 (up to 24 hours!).

The discovery of SN 1987A in the LMC and its progenitor

[Excerpt from *The Astronomer*, Vol. 23, p. 174, 1987. The story begins with Gordon Garradd on his routine Magellanic Cloud nova patrol. [See Gordon's contribution in this chapter.] He photographed the LMC on Feb. 22.499 UT and within 12 hours had checked his photos 'discovering' two faint variable stars. One was a Harvard Cepheid plotted on Mati Morel's *Visual Atlas of the LMC*. The other was checked by myself and found to be another mag. 13 Harvard Cepheid. My own photos were not checked in view of Gordon's searches, but I had reached another stalemate in my searching due to a three-week overdue order of slide mounts. A few slide mounts had been kept aside for mounting my wide-angle and Mag. Cloud shots, but the momentum had been lost and dozens of unsearched films were piling up. Gordon was clouded out on the 23rd, but at 23.44 UT both the Mag. clouds were covered by me from Siding Spring and were developed and mounted within 12 hours. I have often advocated giving cursory scans if time is not available for a complete search. Somehow, on that day, however, I never looked at my photos. It was a long day, as after a terrible nightmare (I haven't had one for several years) which awoke me after a couple of hours, I got up and started several projects which I'd meant to do for some time . . . Several times that day I thought I should check my photos but I never did and the episode is quite mysterious to me. Normally I search without a thought!

Darkness had now descended on Chile, and Ian Shelton was continuing his photography of the LMC as part of his research at the University of Toronto Las Campanas Station. He was taking a three-hour exposure with the (10-inch) astrograph, stopped down to give better quality images, and was battling against high winds as he guided. He had noticed a star in the finder, but not realised that it was unusual until looking at the plate when it was still wet after immediate development. The star was not present on a photo he took from Feb. 23.059–23.191! Trying to be as calm as possible and realising that there were other explanations than a SN [supernova], he went outside and looked at the LMC. IT WAS THERE!, just beside the Tarantula Nebula!!! Again, trying to remain calm he walked across to the 40-inch where he asked if anyone had noticed anything unusual about the LMC. Oscar DuHalde, the night assistant, had noticed something wrong earlier, but had been busy in instrumental changes and had not mentioned it. With confirmation, it was important to get news to the Central Bureau in Cambridge, MA. However, several hours of phone calls failed to get through, and in the morning a message was given to someone going off the mountain into La Serena, the nearby town. Again, the message did not get through, so a Telex message was arranged through the Cerro Tololo Inter-American Observatory in La Serena. At last the word had got through!.

Around this time, Albert Jones, the veteran variable star observer (the most prolific variable star observer of all time) and comet discoverer (1946 VI) had left a committee meeting early as the sky was clearing. Before long, he was observing . . . in the LMC. [See Albert Jones' contribution in this chapter for more details.] Brian Moreno and Stan Walker, two amateurs at Auckland Observatory, confirmed Albert's discovery and measured photoelectrically V = 4.81 . . . at Feb. 24.46 UT. Frank [Bateson] had phoned the AAT, and they had called Tom Cragg who phoned me! I had just started my LMC photos for that night when Tom called!

We confirmed the supernova within seconds and I must admit that, unlike Ian Shelton, I panicked and several times during the night forgot where I had left my opera glasses, torch, etc. My thoughts ran riot, people to call, photos to check, photos to take, plates to measure. First of all, I estimated a mag. of 4.8 on 24.455. Knowing of my Feb. 23.44 photos I was eager to check these, but I had left these at home. This I never did, always carrying my photos for situations like this! . . . Gordon checked his Feb. 22.499 photo again and said that the mag. 12 [progenitor] star was as normal, and there was no new nearby image to mag. 14.5 . . . On heading home, I checked the Feb. 23.44 photos and almost fell over. There was the object at mag. about 6.1!

Amateur searches for minor planets

[Adopted from an article that appeared in *The Astronomer*, September 1989.] In the first 7 months of 1989, 175 minor planets received official numbers. What may surprise readers is that for 11 of these, the principal designations were discoveries of 1989, and seven were discovered by amateurs. These include:

Minor planet	Mag.	Arc	Nights	Opps	Discoverer(s)
(3998) 1989 AB	15	11	3	6	Kojima
(3999) 1989 AL	15.5	8	4	10	Kojima
(4000) 1989 AV	15.5	13	4	6	Ueda and Kaneda
(4042) 1989 AT1	15.5	17	3	3	Endate and Watanabe
(4062) 1989 BF	17.0	36	6	4	Colombini *et al.*
(4063) 1989 CG2	16.5	53	6	4	Colombini *et al.*
(4106) 1989 EW	16.0	23	7	4	Nomura and Kawanishi

Here 'Mag.' = discovery magnitude; 'Arc' = observed arc (in days) over which the object was observed in 1989; 'Nights' = number of nights on which the object was observed and subsequently identified during the 1989 opposition; and 'Opps' = number of earlier oppositions when the object was observed. Note: (4063) is a Trojan asteroid.

The instruments used were as follows:

Kojima 25-cm f/3.4 Wright–Schmidt	Colombini 25-cm f/2.5 Schmidt
Kaneda 16-cm f/3.8 Wright–Schmidt	Nomura 25-cm f/3.4 Schmidt
Watanabe 16-cm f/3.3 reflector	

There were many other Japanese and Italian amateurs involved in discoveries, but the above is representative. It is clear that most discoverers use wide-field instruments, but any instrument which can reach to mag. 17 is a potential discovery instrument.

Observed arcs as little as 5 days have been used at the Minor Planet Center and by Japanese amateurs to make identifications at other oppositions, but observers should aim at a minimum arc of 10 days with observations on three nights. (I should add that

Canadian amateur Andrew Lowe has been very successful in making identifications at other oppositions using only two nights' observations.)

[The above is only a small portion of Rob's excellent article. Readers interested in doing similar work should read the entire article carefully. See Appendix H on how to order copies of this issue of 'The Astronomer'.]

Further comments

[From a letter dated August 20, 1990.] On May 1 I became a 'real' professional astronomer, not one of those upstart geophysicists. Actually, I'm not sure if you know my background. At university, I failed physical sciences and dropped out. After a year of hitch-hiking round the UK I went back to university and did an honours degree in psychology. I have had a strong interest in the evolution and nature of mind, its relationship to the development of science and culture, and more directly the relationships between mind and reality.

For a few years I did a lot of visual patrolling (late 1970s). I've also done a lot of visual comet searching using 20 × 120 binocs. I have tried to encourage amateurs to find minor planets as they are the easiest astronomical objects to discover. So far I have only got Brian Manning in England to try it, and he has around ten new objects since October 1989. As an amateur using professional equipment or going through old and new UKST [1.2-m f/2.5 United Kingdom Schmidt Telescope] plates, I found about 100 new asteroids for which orbits have been calculated. Several have been numbered. This was how I spent my weekends. (People here think I'm a bit strange!) [See Figure 5.16.]

It is a myth that visual discovery is more immediate [than photographic discovery]. This is an *a posteriori* argument! The correct way to look at it is that a *nova* occurs (not a discovery) and it is to be found. If it is mag. 3 or brighter, it will certainly be found more rapidly by a visual searcher. But say mag. 6 or 7 and the visual searcher is in for some hours of searching. A single fish-eye photo would have the discovery in similar or less time. Now go to mag. 8. Very few visual discoveries are this faint (are there any?) and the visual task becomes burdensome, but a standard lens would do the job in around ten photos – and discovery the next day. Mag. 10 and the visual observer has lost outright. Thus, suitably chosen photographic equipment is the key to success. All this said, I must admit that I would get more satisfaction from a visual discovery!

I stopped nova searching sometime in early 1988 and stopped patrol photography in late 1989. Just not enough time to follow both it and my interest in minor planets. Also, my search technique was inadequate and my whole philosophy of search was wrong. I'll only start again when I have a more efficient search method like a quality stereo-comparator for large format films of at least 120 format, preferably bigger. I would in the future concentrate only on the brighter novae. Sure, it's fun finding mag. 10 novae, but I always felt those mag. 6 novae outside the Milky Way

Fig. 5.16. Rob McNaught by the Hewitt Camera at Siding Spring.

deserved to be found as well. Thus, I am having a 35-mm fish-eye lens adapted to a sheet film camera to give an all-sky circular field 9 cm across and thus could search a hemisphere to mag. 8 by only taking and searching one field (two photos).

I found about 20–25 new variables some of which have been published, others to be published soon. Many are Miras [Mira-type variables].

On the night of Feb. 23, 1987, I took many (can't remember how many) photos. I have always been @$#& off with stories of how I 'missed' the discovery [of SN 1987A]. I got the photo because photos were taken often and covering a large fraction of the sky. So I had good coverage of several events and this is why I did it, not solely to make discoveries myself. To miss it is to look for something and not see it – or did everyone (except three people) in the southern hemisphere 'miss' it?

About stereo-comparators: Having used the Shoemakers' and Helin's stereo-comparators, I must say that it is way ahead of a blink comparator. I can't convince Ben Mayer, but I'm very unimpressed with the PROBLICOM [PROjection BLInk COMparator] method and totally disagree that some people can't use stereo [viewers]. I was one of them, but after several hours of effort, got used to it. If someone isn't willing to REALLY TRY at the method for many hours, they can't really be serious about searching. I got fed up with all the PROBLICOM adjustments.

There you have it. I'm really a terribly opinionated person (but with a heart of gold).

[Before you give up entirely on the idea of using a PROBLICOM, you should read Dan Kaiser's contribution. However, I (WL) personally side with Rob McNaught.]

UNTITLED REMARKS
Michael Rudenko, Amherst, Massachusetts, USA

I first came up with the idea of undertaking a comet patrol programme towards the end of 1980 when as many as seven comets were observable in a single night through small telescopes. I simply reasoned that if I could see newly discovered comets, it should be possible for me to discover them. I formally began searching for comets in January of 1981. (My first 70 hours of patrol were with the 23-cm Clark on the roof of the Harvard–Smithsonian Center for Astrophysics building in Cambridge!) I eventually got use of a 13-cm RFT [rich field telescope] which I used to bring out to Harvard's Oak Ridge Station. In 1983 I acquired a 6-inch f/8 apochromatic triplet refractor from Roland Christen, and in 1984 I moved to western Massachusetts whence I discovered comets 1984t, 1987u, and 1989r.

[See Figure 5.17 for a picture of Michael.]

MY LIFE WITH COMETS
Tsutomu Seki, Kochi, Japan

I was astonished to meet the eclipse comet which appeared suddenly in the autumn of 1948. [Comet 1948 XI was discovered by numerous people during a total solar eclipse. Its estimated magnitude at the time was − 3.] That comet had come from deep space giving us only one opportunity to see it. It left the Earth showing a magnificent tail in the dark sky.

After seeing this comet I started to study astronomy, and I especially became interested in researching comets. When I was in high school Comet Honda 1948m was discovered. It impressed me and it was at that time I was moved and wished to discover a new comet.

I started searching for comets earnestly in the summer of 1950. My observation place was the steps of my house in Kochi City. In 1950 the population was about 200 000. My comet seeker was a reflector of 25 magnification with a 4-inch lens which I polished myself. But Mr Leslie Peltier of Ohio, USA, very famous as a comet hunter, and Mr Anton Mrkos of Czechoslovakia, an expert at finding comets, had already found many comets. Therefore, a new face like me had little room to step into the field of finding new comets.

Fig. 5.17. Michael Rudenko (l.) receiving an award (from Steve O'Meara of *Sky & Telescope*) for one of his comet discoveries.

Soon I changed my comet seeker to a reflector with a 6-inch mirror, but I had a lot of trouble bringing it into focus. Over a 10-year period my eyes patrolled the night sky for 1000 hours. During that time I didn't find any comets, but I memorised a lot of stars. I especially memorised the star clusters and nebulae which resemble comets.

As a result of my 10 years of research I detected only one periodic Comet Crommelin having a recurrence cycle of 28 years. But few people know about it.

I had some failures. One winter night I thought I had discovered a bright comet and became very excited, only to become very disappointed when I realised it was the remains of a kite which was hanging on a high voltage electric wire. On another occasion I mistook Mars, which was moving backwards, as a nova and sent a telegram to the observatory of my discovery.

Cold winters bothered me in investigating comets. When the temperature was − 5 degrees centigrade the lens of the telescope became frozen. After continuing my observations for over 3 hours in that temperature my heart and body also became as cold as ice.

Once I gave up trying to find a new comet, but I braced myself again and found Comet Seki 1961f, by using a comet seeker refractor of 15× magnification with a 3.5-inch lens.

Thinking back over the past 40 years I cannot forget my good friendship with Mr Kaoru Ikeya. He and I found a memorable comet, Ikeya–Seki 1965 VIII (see Figure 2.3), in September of 1965. Mr Kaoru Ikeya is very good at making mirrors and he gave me a 4-inch mirror for a reflector in April 1972. Since then the mirror's beautiful face has supported my heart.

I hope to find the 23rd minor planet at the Geisei Observatory and name it Ikeya in honour of Mr Kaoru Ikeya's work finding five comets and some supernovae. By naming a minor planet in honour of Mr Ikeya, I feel our friendship will shine forever.

Fig. 5.18. Tsutomu Seki (centre), Bill Bradfield (left), and Akira Kawazoe (right) at the Geisei Observatory.

[Mr Seki wrote the following to me on May 20, 1990]

Geisei Observatory belongs to the Kochi Education Committee, and they open it to the public one day a week (Monday). I am one of the lecturers there. It takes about one hour from Kochi City to Geisei (about 35 km).

I can use the telescope any time, so I search for comets and minor planets using the 60-cm reflector on fine nights.

[Mr Seki's contribution was kindly translated by Akira Kawazoe. He also took the photograph, Figure 5.18.]

UNTITLED REMARKS
Carolyn S. Shoemaker, Flagstaff, Arizona, USA

We [Carolyn and her husband Eugene M.] not only search for comets and Earth-approaching asteroids now, but we also look for high-inclination asteroids, Mars crossers, and those faraway asteroids, the Trojans. While there are times when Gene and I observe alone, we have added Henry Holt and David Levy to our team, and frequently have one or the other with us on an observing run. This allows me more time to scan films for objects that may need follow-ups while we are still on the telescope.

When Kodak put their Technical Pan 4415 on a 7-mil base in 1987, this film became our choice for work on the 18-inch Schmidt. Because it is a slow film, we hypersensitise it for 6 hours. Now we generally take 8-minute exposures, allowing us to find many fainter objects on the fine-grained film. Our pairs are now usually separated by 45 minutes to an hour – a factor that makes finding slow-moving objects like Trojans easier. We typically take around 40 films a night, or 20 fields; in the course of a year, with good weather, we cover about 40 000 square degrees – equal to virtually the entire sky.

Our observing runs are scheduled for seven nights each month, excepting December. Henry Holt has been taking the lunations during the summer months and will have David Levy with him this year (1990).

Too bad there are not more women amateurs doing astronomical things – if they only knew what fun there is in this field! [Adopted from a letter dated May 19, 1990. See Figure 5.19 for a picture of Carolyn at the Schmidt.]

DISCOVERY OF COMETS IN A MOONLIT SKY
Tetsuo Yanaka, Tochigi Prefecture, Japan

My interest in the stars began when I was in primary school. I used to go fishing in a stream near my house. It was there I first noticed the bright 'Evening Star' in the dusk. Although I was impressed by the beauty of the star, I did not step further into the celestial world at that time.

At the time Mr Honda, Mr Ikeya, and Mr Seki were very much involved in searching the night sky. After a baseball game in my junior high school, I asked a senior student about that Evening Star, and he told me what it was.

Soon afterwards I bought a small telescope. I was fascinated when I first looked at the surface of the Moon. This gave me the chance to step into the world of astronomy.

After I entered senior high school, I earned some money by delivering newspapers, and I bought a 10-cm reflector with a German-type mounting. I started to study astro-photography and to observe planets and meteors. In January 1970 I observed Comet Tago–Sato–Kosaka [1969 IX, discovered in October 1969] every night. It had come up from the south and passed to the north. From then on my heart was entirely captured by the sky walkers, comets.

It was on February 1, 1970, that I started a full-scale search for comets. I remember it was a very cold morning, and I searched in Aquila using the reflector which I had remodelled. Soon I found a nebulous object of 7th magnitude in the morning twilight. No such object was on my star map. I sketched it because I thought it might be a comet, but the day soon dawned.

As I was a beginner, I thought I would make a report to the observatory the next morning having confirmed the comet's movement. But on the same morning I learned of the discovery of a new comet named Comet Daido–Fujikawa [1970 I, discovered on January 26].

Fig. 5.19. Carolyn Shoemaker and the Palomar 46-cm Schmidt.

As I had heard stories about how other well-known comet searchers had similar experiences, I didn't lose heart. Thinking that I was young [Mr Yanaka was born in 1954], I felt I could find many comets in the future. But in retrospect I now realize that my way of thinking 20 years ago was perhaps too optimistic.

In those days I searched for comets as hard as I could, but no comet came. When I was discouraged and gave up my work, comets would ironically appear. Anyway my opportunities for searching were sporadic. During that time many comets appeared. I remember especially the brilliance of Comet Bennett and Comet West even now.

A few years later my father became seriously ill, and since I was young I began to worry about my own health. I cut back my searching in order not to overwork myself. As a result, my observations became less frequent.

I suffered in this pitiful condition for a while. I knew it was important to take care of myself, but I wondered if I really had to give up my dream. A man can live at most for a hundred years, but this is certainly very short when compared with the life of stars. A short and dreamful life would be superior to a long monotonous one. And the older I become, the more exhausting the observations will be. So I chose to pursue my dream no matter what.

In 1985 I completed my long-wanted observatory – the 3-m diameter dome became my own spaceship, and I started to search for comets again.

I found Comet Machholz (1988e) independently, and it gave me much confidence; I knew definitely I could find my own comet. Later I bought a pair of 15-cm binoculars and tried to search very carefully. And I always imagined a scene that a comet would appear in the field of vision. And then after my heart would calm down, I would regain my composure and again concentrate on searching for comets.

At 5:07 a.m. on the morning on December 30, 1988, an object like a double star of the 9th magnitude came into view. My mind wavered for a moment whether to pass it or not. But I decided to look at it with my 40-cm reflector. There I saw that it was a nebulous object with a central condensation, not a double star at all. I was trembling with excitement as I knew there was no nebula of the 9th magnitude just to the west of M12 [a 6th magnitude globular cluster in Ophiuchus]. I realised that it was surely a new comet.

Immediately I sketched it and confirmed its movement using the 40-cm telescope again. I called to the National Observatory in Tokyo and Mr Levy in America confirmed it. And Yanaka's comet (1988r) was born.

Just three days later, in the morning of January 2nd of the new year, I picked up a faint object of the 11th magnitude near the boundary between Bootes and Virgo. At first I thought it was again a double star, but when I focussed more carefully, it turned out to be a single star.

I thought it seemed interesting and looked at it through the 40-cm reflector. And then I caught sight of another nebulous object right next to the faint star and partially overlapping it. After sketching it, I watched it some more and to my second surprise, the two objects separated.

I was wondering if it was possible that I had found two new comets in such a short time! I pinched my cheek to make sure it was not a dream. The comet was soon confirmed by Mr Koishikawa at Sendai Observatory and by Mr Morris in America. It was named Yanaka's comet (1989a).

I still wonder how I could have made both these discoveries under the unfavourable condition of considerable moonlight. [Last quarter Moon had occurred on

December 31st. See Figure F8 in Appendix F.] Maybe they would have gone unnoticed if I hadn't been so enthusiastic.

Lastly, I would like to express my sincere gratitude to my family who have understood and supported my hobby of comet-seeking. There are many people who are devoting themselves to comets. I hope Heaven grants them happy light! because up to this point I had been frustrated in my endeavours to find comets.

[The above version of Mr Yanaka's contribution is the amalgamation of two excellent translations, one by Toru Hayashi and the other by Akira Kawazoe.]

SHORT BIOGRAPHICAL NOTES ON THE CONTRIBUTORS

George E. D. Alcock

Born 1912, Peterborough, England. Although he claims to be much more a meteorologist than astronomer, his records for 1989 reveal that during that year he spent 293 hours and 23 minutes sweeping the skies. At 4:35 UT on March 25, 1991, a half hour after the sky cleared off and as the dawn was rapidly brightening, he discovered Nova Herculis. At the time, he was inside his house using 10×50 binoculars and looking through a large double-glazed window. It was the 12th night of that month that he had spent time searching, and his fifth nova discovery, not including finding RS Oph in outburst in 1985. He also has five comets to his credit. (Notes adopted from *The Astronomer*, Vol. 27., No. 324.)

Rodney Richard Dacre Austin

Born 1945, Christchurch, New Zealand. Royal New Zealand Air Force, 1963–66. 'Discovered' Comet Honda 1968c two months too late. (See note on David Levy.) Joined staff of Mt John University Observatory (NZ) in 1971. (Resigned for reasons of health in 1979.) Now works on the night shift (6:30 p.m. – 2:30 a.m.) of Taranaki Newspapers Ltd, New Plymouth, NZ. WL first met Rod at a meeting on Asteroids in Tucson, Arizona, in 1979.

Reverend Kenneth Beckmann

Currently Pastor of the Congregational United Church of Christ in Lewiston, a small resort community in the northeast portion of the lower peninsula of the state of Michigan, USA. Author of the current edition of *A Nova Hunter's Handbook*, published by the AAVSO.

William Ashley Bradfield

Born 1927 in Levin, New Zealand, son of a dairy farmer. Degree in mechanical engineering from the University of New Zealand. Left NZ in 1951 to work for the

Australian Government on research and development of solid propellant rocket motors for upper atmosphere research vehicles launched from Woomera. After 10 years as Principal Research Scientist in charge of a rocket propulsion group, he retired in 1986. Active in the Astronomical Society of South Australia. Has discovered more comets (15 as of February, 1992) in the 20th century than any other amateur.

Michael J. Collins

Lives 70 km north of London in a 'rather poorly lit town close to the Great North Road of old'. During daylight hours he works for the Institution of Electrical Engineers in charge of a small group of information scientists abstracting articles for the INSPEC Database (available on-line throughout the world) and for *Physics Abstracts*.

Peter L. Collins

Born in 1949 and attended Harvard College but did not obtain a degree. He recently moved to Boulder, Colorado (from Tucson, Arizona) where he says the sky is darker. Unfortunately, from his home there is a mountain blocking the view of Sagittarius, but he plans to set up his observation post where he can see the greater part of this nova-rich constellation.

Reverend Robert Evans

Presently Minister of the Central Blue Mountains Parish of the Uniting Church in Australia, located in Hazelbrook, a small town 80 km due west of Sydney. (Note: Bob will surely be embarrassed to know that he apparently did not see the suggestion that I made (to all contributors) that they limit their remarks to five pages. Being that Bob has made a truly unique contribution to visual discovery of supernovae, I have made no effort to abridge his remarks.)

Gordon Garradd

Born 1959, now lives in Tamworth, NSW, about 320 km due north of Sydney, Australia. Gordon is currently engaged in a friendly competition with WL to find novae – or another supernova – in the Magellanic Clouds.

Toru Hayashi

(Translator of the contribution by Honda, and co-translator of the contribution by Yanaka.) An archaeologist born in 1950. Now lives in Tokyo. Was an exchange high-school student in California for two years. Has worked with WL on archaeo-astronomy in Easter Island.

Eleanor F. Helin

Works for the Jet Propulsion Laboratory of the California Institute of Technology, Pasadena. WL first had the pleasure of meeting Eleanor on the same day he first met Rod Austin (see above).

Minoru Honda

Born in 1913 in Hatto-cho, Tottori Prefecture. Agriculturist, lived in Kurashiki-shi, Okayama, about 120 km north of Hiroshima in the southwestern peninsula of Honshu, Japan. His first discovery, in September 1940, was Comet Okabayashi–Honda 1940 III. Died (of cardiac insufficiency) on August 26, 1990. (I received Mr Honda's contribution on August 10, 1990.) Mr Honda's obituary records that 'In the 5th grade of elementary school, he started watching the stars in a box wearing the huton quilt.' The translator, Toru Hayashi, tells me that the 'huton' is a traditional Japanese bedcloth which still is popular today. Honda is survived by his widow whose name, Satoru, in Chinese characters is the combination of 'comet' and 'heart'.

Albert Jones

Born Christchurch, New Zealand, 1920. A high-school graduate, he first began reporting aurorae australis in 1939. Joined the New Zealand Astronomical Society in 1941. Has won numerous awards from prestigious groups such as the Royal Astronomical Society of London, the British Astronomical Association, and the Astronomical Society of the Pacific. In August 1987 he was awarded the Order of the British Empire (he was nominated *before* his independent discovery of SN 1987A).

Daniel H. Kaiser

Born 1948, interested in astronomy since 8 or 9 years old. Served in the US Army as a helicopter pilot from 1967 until disabled. Bachelor's degree in business administration from Ball State University, Indiana. Now operates Kaiser's Shoe Repair (since 1979). Perhaps the most successful user of the original PROBLICOM concept.

Akira Kawazoe

(Translator of the article by Mr Seki and co-translator of the article by Mr Yanaka.) Born 1934. Lives in the city of Kochi on the island of Shikoku, Japan, and teaches high-school natural science (mainly geology and astronomy). Has travelled widely (USA, Canada, Iceland, Africa, Australia).

David H. Levy

Born in Montreal, Canada. Degree in English literature from University of Toronto. 'Discovered' Comet Honda 1968c two months late. (See note on Rod Austin.) Writes 'Star Trails' in *Sky & Telescope*, and 'Observer's Cage' in the Newsletter of the Royal Astronomical Society of Canada. Author of *Observing Variable Stars* (see Appendix H). WL first met David at a conference on Halley's Comet in Heidelberg, Germany, in October 1986.

William Liller (WL)

Born in Philadelphia, USA in 1927. Fascinated with the heavens since seeing a solar eclipse in 1932. Received degrees in astronomy from Harvard University and the University of Michigan, and afterwards was a professor at both these institutions. Took early retirement and became a born-again amateur in 1983 when he moved to Chile. Co-authored (with Ben Mayer) *The Cambridge Astronomy Guide*.

Donald Edward Machholz, Jr

Born 1952 in Virginia, USA, degrees in laser technology and general education. Interested in astronomy since age 8. Currently is the ALPO Comet Recorder. Also writes *Comet Comments*, a two-page monthly newsletter. [Highly recommended – WL]

Brian G. W. Manning

Born in Birmingham, England, in 1926. Made his first mirror from a piece of glass that a World War II bomb blew out of the roof of the factory where his father worked. Engineering draughtsman, later metrologist at the University of Birmingham. Early retirement in time for Halley's Comet. In the late 1950s constructed interference-controlled ruling engine in home workshop which ruled high-quality 3 by 2 inch gratings. As interested in making optics, mechanics, and electronics as observing. Contributes astrometric measurements of minor planets Galileo, Gaspra, and Ida to NASA for spacecraft fly-by. In 1990 received the H. E. Dall prize of the BAA.

Robert H. McNaught

In 1984 Rob went to Australia from England where he had been an active amateur astronomer. Has worked ever since at the Siding Spring Observatory in Coonabarabran, NSW, Australia, first with the UK Earth Satellite Research Unit, and now on the Apollo asteroid search programme of the University of Adelaide. Preferring to telephone rather than write letters, Rob has rung up WL on a number of occasions to confer on recent discoveries (his or mine).

Michael Rudenko

Born in 1955, Bachelor's degree in mathematics from Massachusetts Institute of Technology. Worked as computer specialist at the Harvard–Smithsonian Center for Astrophysics (after WL retired and moved to Chile). Presently employed at the University of Massachusetts in the design and implementation of a new high-performance computer.

Tsutomu Seki

Born in 1929. Now lives in the south of Japan in the city of Kochi on the island of Shikoku. The leading all-round amateur astronomer of the world, according to Brian Marsden.

Carolyn S. Shoemaker

Works for the US Department of the Interior, Geology Survey, in Flagstaff, Arizona. Married to astro-geologist Eugene Shoemaker. The Shoemakers have discovered more comets in the 20th century than anyone else (16 as of early 1990).

Tetsuo Yanaka

Lives and works as a postman in Motegi in the prefecture of Tochigi about 100 km north of Tokyo.

6

Seeing it first: visual searches

Chapter 5 contributor David Levy gave probably the best reason I know to use visual searches to look for things in the sky: it gives the observer a chance to see 'live' the many astronomical marvels. Actually, he put it the other way around: the main reason that he takes his telescope out at night is to enjoy viewing the incredibly fantastic cosmos. If something new turns up in one of his sweeps across the sky, fine. But if not, also fine. It's a no-lose situation.

Asteroids are just about the only kind of discoverable phenomena almost never found visually nowadays, and even these solar system beasties occasionally turn up on AAVSO variable star charts and the Thompson and Bryan supernova search charts. Meteor shower and fireball patrols require virtually no equipment at all, of course, and in fact one astronomer, who shall not remain nameless, my good friend Arthur Hoag, claimed that he preferred to close his eyes while watching for fireballs: in that way he would filter out the faint meteors. That's what he told me.

In this chapter we begin by considering, but only briefly, the various kinds of optical equipment and mountings that are available for visual sky patrols. Further details can be found in any number of excellent books and magazine articles. (See Appendix H.) Then we will get down to the mechanics of sky searching, recalling some of the words of the previous chapter, but also adding new ones gleaned from the literature or from personal experience.

Telescopes
Magnification

One of the ways of calculating the magnification of a visual telescope is to divide the diameter of the objective by the diameter of the beam of light that emerges from the eyepiece, the exit pupil. Simple enough. But for every telescope, there is a minimum desirable magnification since all the light that exits from the telescope has to squeeze through the iris of the observer's eye. (See Figure 6.1.) At night fully dilated pupils can range in diameter from 5 to 8 mm. This means, therefore, that the magnification must be no less than one-fifth to one-eighth the diameter of the objective expressed in millimetres. Thus, a 150-mm diameter objective should not be used at powers much less than 30 or else light will be wasted.

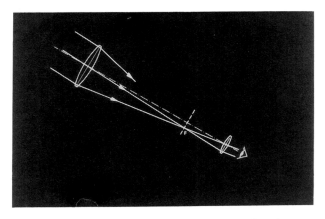

Fig. 6.1. A schematic diagram of a refracting telescope and an observer's eye.

The upper limit of magnification is generally determined not by the telescope, but by the steadiness of the atmosphere: 400 or 500 power is about all that one can usually usefully use. If you have one or more eyepieces that give much more magnification than that, save them to use with a shorter focal length telescope – unless you're planning to go into orbit.

Field of view

Those who seek comets and hunt novae will naturally want a telescope or binocular with a wide angular field of view. Thus, the telescope objective must condense the imaged sky down into as small an area as practicable, and that calls for a short focal length. But since one nearly always wants a large aperture, the combination of large objective diameter and short focal length combines to dictate a small focal ratio.

With a small focal length objective, light rays coming from the edge of the objective will be refracted (or reflected) at a steep angle. The eyepiece has to be able to accept all the light arriving from the objective and render the rays from each celestial body suitably parallel for the observer's eye. In other words, the design of the eyepiece is critical in the matter. Many visual comet hunters use Plössl or orthoscopic eyepieces; they are relatively expensive but well worth it.

There are also limitations on the focal ratio of the objective, refractor or reflector. Large objectives with small focal ratios – less than about f/4 or f/3 – produce serious aberrations as soon as an object is moved away from the centre of the field. To be sure, camera lenses do well at f/2 or even f/1.4, but they contain numerous elements, some made of exotic glass. Only because the lenses are relatively small can prices be kept within reason.

Limiting magnitude

How faint can you see with a given telescope? Many factors are involved: the size of the objective, quality of the sky, magnification, type of eyepiece, type of telescope, and the physiology and experience of the observer. But see Figure D1 (Appendix D) for a starter.

The *light-gathering power* of a telescope goes as the area, or diameter squared, of the objective lens or mirror. Obviously, the bigger, the brighter, at least in principle. Now let us consider the other factors.

The *atmosphere* is a complex optical medium through which to look; its effect on limiting magnitude can vary enormously from night to night. On a superbly clear night from sea-level, about a quarter magnitude of visual light is extinguished as it passes through the air above us. But away from the zenith, this extinction is greater. Still, 1st magnitude stars (and comets) can be seen almost right down to the horizon.

However, on a 'fairly good' night when haze reduces the brightness of a zenith star by, say, one magnitude, observations near the horizon are virtually hopeless because at an altitude of 30° above the horizon, we must look through twice as much atmosphere as at the zenith and the loss amounts to 2.5 magnitudes; at altitude 14.5°, we must contend with four air masses or a 5-magnitude loss. Lower than that, the opacity of the atmosphere increases very rapidly. (See Figure 6.2.)

On a putrid night, you should work the zenith region and nothing else.

To put it more precisely (but not quite exactly), air mass increases as the secant (1/cosine) of the zenith angle. Experienced comet hunters who often work close to the horizon know well these vicious facts of light.

Seeing, that ill-defined word used to describe the steadiness of the atmosphere, can make a well-focussed star image appear almost point-like, or wash it out into a fuzzy blob many seconds of arc in diameter. When an object is faint, the eye has to pick out an image immersed in the light of the background sky; it obviously will do better when an image like that of a supernova in a bright-armed spiral is sharp. With bad seeing, light loss can amount to many tenths of a magnitude. And when the sky is bright because of nearby city lights or moonlight or an aurora, the losses can be devastating, especially for diffuse objects like comets. There is, however, some salvation, short of moving out into the country as many of the top comet and supernova hunters do. See under 'Filters', below.

Magnification comes into the picture because light entering the eye near the edge of the iris is less efficiently registered. Thus, the higher the magnification, the more efficient is the eye, at least up to the point where star images become, in effect, extended sources. To give a concrete example, a 25-cm objective at 300 power will reach stars a magnitude fainter than the same objective at 60 power. (See Figure D1.) For comets the situation is more complicated since they can appear in various guises: some comets are almost stellar in appearance, while others are diffuse and illusive blobs that barely exceed the background sky brightness.

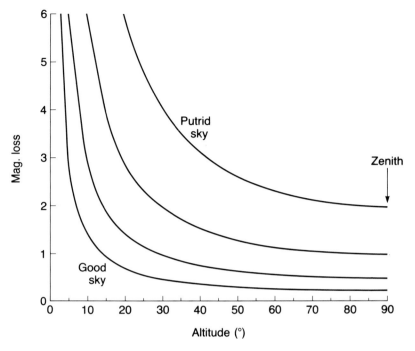

Fig. 6.2. The Earth's atmosphere as an optical medium.

Technically superb *eyepieces* like Plössls and orthoscopics, composed of four or five or more separate lenses, can contribute significantly to light loss if the air-to-glass surfaces have no anti-reflection coatings. So if your favourite ocular is a pre-WWII antique or has been badly treated, you will gain as much as a half magnitude by replacing it with a quality eyepiece with expertly coated lenses. And if you wear glasses, or use a special filter (see below), or observe through your bedroom window, you will be losing more than a tenth of a magnitude per component.

The same words apply to the components (usually two) of a refractor objective, but that does not necessarily mean that refractors are inferior to reflectors. Remember that most reflecting telescopes have a central obstruction – a secondary mirror – that can block out several tenths, or more, of a magnitude. And perhaps worse, it scatters light.

Finally, we must add to these detractors the fact that *visual acuity* varies considerably. Some people can simply see fainter than others. But it does not depend solely on retinal sensitivity; much has to do with experience and patience. It has been estimated that the difference between a 6-second observation and a 60-second observation corresponds to about half a magnitude. One should wait patiently for the next possible flicker of light.

Binoculars

During World War II, it became vitally important to be able to see clearly under low light conditions, and that meant binoculars with large objective lenses. And so they grew. Starting with the popular and convenient 6 × 30s (the first number is the linear magnification; the second is the lens aperture in millimetres), military binoculars began to grow reaching as large as 25 × 150, and perhaps even larger. Eventually, some of these giants came on the surplus market and many amateurs snapped them up. Later amateurs began building their own. (See Figure 5.12.) The largest that I know of is a monster with twin 61-cm reflecting telescopes mounted so that the two foci can be brought together and viewed comfortably with both eyes at the same time.

Having two identical optical systems feeding the two eyes separately is complicated by the fact that human eyes are usually from five to seven centimetres apart. Hence, there is the need − at least with objectives larger than this dimension − to insert prisms or mirrors in the two light paths in such a way as to bring them close together, and that means further light loss − and higher cost.

And so why binoculars? One obvious reason is comfort: holding one eye shut for long periods of time, or using eye patches or light shields, can be uncomfortable or inconvenient. A more important reason is that two eyes working together can see fainter than one; best estimates and theory suggest 0.4 magnitudes fainter. But of course this improvement could be overcome in a monocular by using an objective whose diameter is 19 per cent larger. At current prices, this corresponds to an increase in expense of about 50 per cent, a good deal less than the increased cost of equal-sized binoculars.

Of course, for some tasks, like looking for novae in the Milky Way, a pair of binoculars is sufficient; too much light-gathering power can be overwhelming and bring more stars into focus than anyone could ever hope to memorise. Kenneth Beckmann is completely happy with his 10 × 50s.

As for comet hunting, large binoculars have proved their value in recent years. Since 1985, at least six discoveries have been with this kind of optics: two by Machholz with home-assembled 27 × 127s (at a cost of less than $400!), two by Yanaka with 25 × 150s (within three nights!), one by Terasako (20 × 120s), and one by Ichimura (120-mm binoculars, no magnification given).

Mountings

Many types of mountings exist, but most importantly, telescopes and large binoculars need to be mounted firmly; the use determines the best type of mount, equatorial or alt-azimuth.

At least for a beginner, racing from galaxy to galaxy à la Bob Evans almost demands a motor-driven equatorial mount with accurate setting circles or a digital coordinate read-out system. The latter have become so sophisticated that accurate polar adjustment is no longer needed, and their ease of use is unquestioned. But as the good Reverend has demonstrated ever so convincingly, these handy commodities are only conveniences. Star- or galaxy-hopping works just fine for him; it just takes practice.

Of the several different kinds of equatorial mounts commonly used, each has its advantages and disadvantages: fork mounts are easily balanced but difficult (or impossible) to use on the celestial meridian near the north and south horizons; the German-type 'cross-arm' mounting is relatively easy for the amateur craftsman to build but requires a heavy counter-weight accurately positioned. Smoothly turning Dobsonians can be made to track equatorially but are somewhat complicated. Which of these is best again is determined by the use to which the telscope will be put, and partly by the observer's personal preference.

Most successful comet hunters use alt-azimuths of one sort or another since, as we will see, the preferred means of sky sweeping is moving the telescope either parallel or perpendicular to the horizon. Much of the reason for this preference has to do with comfort: the observer's eye should ideally remain at the same level above ground during each sweep. How this can be accomplished can be seen by studying the telescope mounts used by successful comet hunters. See especially the photographs accompanying the Chapter 5 contributions by Rod Austin and Bill Bradfield.

The current popularity of the Dobsonian alt-azimuth mount arises from its ease and low cost of construction and the magic of teflon which allows the telescope to swing smoothly with just the right amount of friction and negligible backlash. A telescope tube counter-weighted so that the eyepiece stays close to the horizontal axis makes a fine comet seeker.

The main disadvantage of using an 'alt-az' for comet hunting is that when a candidate turns up, locating it on a star map can be frustrating if the region of the sky has few bright, recognisable stars. But modern technology has come to the rescue: digital setting circles that indicate where you are immediately can be installed on your Dobsonian or whatever kind of alt-az you have. The current cost: under $500.

For watching variable stars, planets, and the Sun, the disadvantage of alt-azimuths should be obvious: to follow the object for more than a few minutes requires having to move the telescope in two coordinates. But this certainly is a non-fatal inconvenience. The most novel mount of all those described in the previous chapter is, without question, Albert Jones': he uses an equatorial but chooses the polar direction as it suits him. What a splendid idea!

Filters

At my urban area home, the faintest zenith star that I have ever seen on a first-rate night is magnitude 5.4, and I have reasonably sensitive eyes. The problem is light pollution; my magnitude limit is set by the brightness of the night sky.

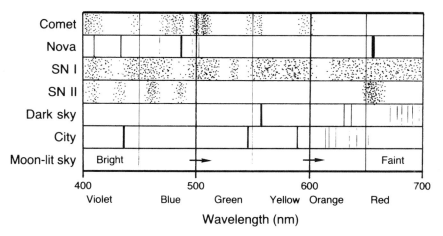

Fig. 6.3. A spectral diagram showing where the strongest emission features appear. City lights are assumed to be a mixture of mercury and sodium radiations. The emission features are shown as on a negative (i.e. dark).

Can filters help? There are two separate matters to consider: the nature of the source of the sky brightness, and the nature of the object itself. First, about the sky. Moonlight is reflected sunlight, and therefore, when the Moon is bright, the colour of a cloudless night sky is blue, just like in the daytime – although the light intensity is too low to make this obvious. Therefore, a filter that blocks out blue and violet light – for example, a light yellow filter like Kodak's Wratten 4 or Wratten 8 – cuts down the sky brightness appreciably without diminishing the brightness of stars and comets too much. (See Figure 6.3.)

When the sky is lit up by grim violet-green mercury lamps as mine is, the same Wratten filters will eliminate the violet emission, but to get rid of the green component, one has to go to deep yellow or orange filters. However, there is now the sensitivity of the eye to consider: its night sensitivity to orange and red light is low. A special LUMICON 'Deep Sky Filter' (see Appendix J) that eliminates everything from the mercury violet and to the sodium yellow is available, but is probably better suited for photography.

As for the light scattered from harsh-yellow sodium lamps, it is possible, in principle, to make a filter that blocks out just the relatively narrow band of colour (especially in low-pressure sodium lamps), but it would have to be ordered specially and would be expensive. For visual work, sodium lamps are bad news.

But there is some good news. Comets that have begun to sprout a coma (see Chapter 2) will glow at certain distinctive wavelengths. As a comet approaches the Sun, one of the first emissions to show up comes from carbon vapour with the strongest of several colour bands (called Swan bands) appearing with a pale green light. These developing comets can then be further enhanced relative to the background by special carbon light filters. LUMICON sells one (see Appendix J), and a Wratten 60 also helps.

Shortly after they erupt, novae and supernovae, especially Type II, shine especially brightly in the fairly deep red (hydrogen alpha) owing to the presence of hot hydrogen and ionised nitrogen. Furthermore, if semi-hidden behind an interstellar dust cloud, any star will look red. However, the eye's sensitivity is low in this extreme colour, but if you have ample aperture, you might do well using a deep red filter, like the Wratten 29, to make the nova or (Type II) supernova stand out. Certainly, once you have made a discovery, you may find that a red filter will suppress all other stars far more than your newly found nova or supernova.

In short, filters can help, but alas, not as dramatically visually as they can photographically. (See next chapter.) Some commercially made eyepieces come with a threaded ring into which a standard glass filter can be screwed. Failing this convenience, buy a selection of Wratten gelatin filters and cut them to size, ready to mount in a spare eyepiece (preferably inside to keep moisture off).

Scanning the sky

Basically, the search procedure involves scanning across each field slowly enough so that the eye–brain combination will work at its highest efficiency to pick out the searched-for object, whether it be a fuzzy patch or a star that normally is not there. How much time one spends on a field depends on how close to the magnitude limit one wants to go. As was noted earlier, the difference between a 6-second observation and a 60-second observation corresponds to roughly a half magnitude. And as Reverend Bob Evans pondered, 'I have often thought of the many times I have glanced quickly at a galaxy without taking such a concentrated look, and I have wondered what [SNe] I might have missed as a result.' (Chapter 5.)

I once asked Rod Austin to estimate how long he spent on each field, and he replied, 'I try to move along quickly, spending only a second or two on each field.' He surely would have added that he moves along most quickly when there are few stars in the field. It would seem that Rod's success comes from covering lots of sky, but not down to the limit of his telescope.

Asteroids have to be the biggest nuisance of all in nova and supernova hunting, especially when searching near the ecliptic. In visual nova hunting where the magnitude limit is around 9, the simplest way to verify a new find is to keep informed on what asteroids are around. The several annual handbooks (BAA and RASC, for example) and especially the 'Minor Planet Observer' are good sources of information. When the magnitude limit is much deeper, as it is in supernova searches, then the passage of time is the best discriminator. A typical asteroid will move a second of arc in two or three minutes; few would escape detection after a half hour.

Comet seekers must contend with galaxies and nebulae, some of which can look disarmingly similar to comets. Searching in the Virgo Cluster is especially trying for the visual observer; it harbours over 500 galaxies brighter than 15th magnitude. The best weapon is a good memory so that precious time is not wasted checking out each of the galaxies that lurk there. Here comparison photography (see next chapter) has a great advantage.

By now most of the above remarks should be merely reminders. The experts have spoken (in the previous chapter); prospective discoverers would do well to read and re-read their words carefully.

7

Catching it on film: photographic patrols

A thousand words? A photograph is certainly worth far more if it's astronomical. It provides a permanent record of often many tens of thousands of stars (some variable), plus galaxies, nebulae, and (perhaps) comets, asteroids, novae, supernovae, and meteors that can be measured and analysed in a myriad of ways; it can be printed or processed with revealing techniques that bring out subtleties not apparent to the casual looker; it can be shown or mailed or transmitted electronically to others not conveniently nearby; and it's cheap and easily made. (See Figure 7.1.)

Six years ago when I bought the simplest (and cheapest) available Nikon body and a f/1.4, 85-mm lens in a duty-free shop for less than $500, I really had no idea as to how efficient a nova-finder the combination would be.

This chapter, devoted to photographic patrols of the sky, will draw upon much of the information given in the previous chapter since the eye's retina, like the photographic emulsion, is really nothing more than a sensitive area detector. The eye has its limitations, but its biggest problem has nothing to do with the eye itself: the problem lies with its information processor: its memory is faulty, and it sometimes interprets the images it receives in strange and unusual ways. As we will see, photographs do, on occasion, lie, but in general they are far more reliable as data banks than the human brain.

Nikons, etc.

So much has been written about ordinary cameras that little else needs to be said. For astronomical discovery, one wants a fast lens (preferably f/2 or faster) to keep exposures short, and one chooses the focal length to fit the task. Aside from these considerations, all that one really has to look for in a camera is the means to make a time exposure, usually accomplished nowadays by the B (for 'Bulb') setting, and a cable release that can lock open the shutter for the desired period of time. The image quality produced by cheap lenses may leave something to be desired, but the fierce competition in the camera industry keeps the quality impressively high.

Wide-angle lenses with focal lengths less than 50-mm are excellent for recording huge chunks of the Milky Way where the novae lurk. Something between 50 mm and 150 mm allows one to zero in on specific regions of the sky – for example, a narrow belt around the Milky Way, or that richly endowed area in Sagittarius where

Fig. 7.1. Comet Liller 1988a four months after discovery. Photograph taken by John Sanford, Anza OCA Observatory.

roughly a third of all galactic novae appear. Also, telephoto lenses serve well for reaching to a fainter magnitude (see below), such as might be wanted for comet patrols.

In Appendix D you will find a handy list (Table D2) giving, for each of the more popular lens sizes that is commercially available, useful specifications such as focal scale and size of the field.

Telescopes

At focal lengths much longer than several hundred millimetres, standard lenses give way to common refractors and reflectors that can provide a suitable scale when looking for supernovae partially hidden in the confusion of a bright spiral galaxy. Any decent telescope can be turned into a camera by putting a piece of film either directly at the focus of the objective, or behind an eyepiece adjusted to produce a magnified second focus. On the commercial market one can find a variety of devices made for the purpose, often for specific telescopes and with built-in guiding eyepieces. Most centre around an ordinary camera body so that 35-mm film can be used.

For the majority of galaxies, photographs of excellent quality and well-suited for hunting supernovae are taken at the direct focus where the scale will be close to ideal. Again f-ratio is an important factor – to keep exposure times at a minimum – and telescope objectives of f/4 or f/5 perform admirably. Later we will suggest routines for accumulating photographs of galaxies which may just be harbouring a supernova or two.

For planetary photography, eyepiece projection is, for most, the preferred method. Again a commercial camera body, preferably a single-lens reflex type, is mounted behind an eyepiece racked to give a sharp focus at the film surface.

The Sun's brightness presents both problems and advantages for photography of this nearest star: on the one hand, it is necessary to reduce greatly its intensity, but then on the other, short exposures are still possible. One of the best ways to accomplish both is to place an inexpensive and readily purchasable aluminium-coated mylar sheet over the objective. Another eminently simple method is eyepiece projection onto a white sheet of paper. When mounted several dozen centimetres away, this screen can be photographed directly with any commercial camera.

Pure Schmidts

Standard telephoto lenses with focal lengths greater than 300 mm are simply not available at f-ratios less than f/4 or so. When one wants to photograph a reasonably large field of view with short exposures, Schmidt optics are the order of the night.

Probably the most popular Schmidt camera manufactured today is the 8-inch (20-cm) f/1.5 Celestron, although other similar instruments can occasionally be found advertised in the pages of semi-popular astronomy magazines. The Celestron optics are excellent and capable of producing star images no larger than the finest photographic emulsion grain over a field that measures 4.3° by 6.3°. The critical distance from mirror to focal surface is maintained by low-thermal-coefficient Invar rods, and the focus is carefully set at the factory. A piece of standard 35-mm film is cut to a 40 or 45 mm length and inserted in a film holder that presses the film to the properly curved focal surface. One then opens a small door at the side of the tube and installs the film holder which is securely held in place magnetically. It all works like a charm. The only occasional problem results when the film buckles ever so slightly, or a tiny grain of grit or dust lodges between film and film holder and the emulsion is thrown out of focus. At f/1.5, the focus is extremely critical.

Larger and smaller aperture Schmidts are available with similar f-ratios; again, one has to match the optics to the purpose.

Mountings

Except for solar and some planetary photography, motor-driven equatorial mounts are an absolute necessity – and plan on doing some guiding. If you are shooting for novae and already have an equatorially mounted telescope, you can easily mount your hand camera piggy-back and guide with the same telescope. If you have no such telescopic equipment, you can buy or build a small equatorial mount, and mount the camera together with a small (50-mm aperture) guide telescope. It's really quite easy. Several companies, listed in Appendix J, sell mounts just for small cameras.

Different types of equatorial mountings and their relative advantages and disadvantages were discussed in the previous chapter. For photography at or near the eyepiece, about the only word of caution to be added is that fork mountings can become awkward (or useless) near the horizon on the meridian. Consider carefully the overall system before buying or building.

Film

Emulsion characteristics vary greatly, but most are designed to be used at short exposures in bright light, like the various 'chromes' and print films and popular black-and-white emulsions like Tri-X and T-Max. Fortunately, in recent years the properties of these films have improved to the point where most are excellent for astronomical use. However, a few technical points are important to note.

Film speeds, expressed as ASA or ISO numbers (proportional to the sensitivity), are now available up into the tens of thousands, meaning that in principle it is possible to take decent sky photographs in a matter of seconds. However, the general rule continues to be that the higher the film speed, the more 'grainy' your photographs will appear. (These grain clumps can be easily seen with a 15- or 20-power magnifier or microscope – and in Figure 7.2.) Try as they will, film laboratories have met only limited success in keeping the granularity in check as they develop faster films. Kodak's T-Max is advertised as having grains re-shaped into a 'tabular form' with more surface to catch light, which allows films with extremely fine grain to be made faster. A step in the right direction! Astronomically, the size of the emulsion grain is of tremendous importance since star images or details on the planets are of miniscule dimensions. Only if you were shooting for diffuse comets and nothing else might you be able to get away with super-high-speed film. Table

135

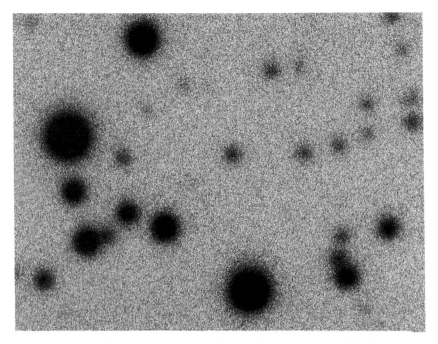

Fig. 7.2. Under high magnitude, photographic grain becomes apparent. The faintest stars
are just smudges in the background 'noise'.

D1 in Appendix D lists a number of the more important characteristics of the most-
used photographic emulsions.

Kodak remains the astro-photographer's best friend. Its publication P-315,
'Scientific Imaging with Kodak Films and Plates', gives a condensed but remarkably
complete course on photographic theory, and I recommend it to any serious astro-
photographer. Kodak, of course, manufactures a wide variety of emulsions, some of
which are mentioned here and in Appendix D. Its 'spectroscopic' plates and films
have been used by professionals for years, and some are available through LUMI-
CON and photo mail order houses.

Hypersensitisation

One straightforward solution to the general rule of 'fine grain means slow film' is to
use a slow, fine-grained film and then increase its speed by means of one of several
special treatments before or during exposing. One common method is keeping the
film at very low temperature with a 'cold camera' during exposure, but this requires
using dry or liquid nitrogen ice and a thermal window to prevent condensation from
forming on the film.

Fig. 7.3. LUMICON's most popular film hypersensitisation kit: an easy way to bring 2415 TP film up to convenient speeds.

Slightly complicated to prepare but far, far more convenient to use is film baked for a few hours up to a few days at a temperature of 50 or 60 °C in an atmosphere containing hydrogen gas. (Pure hydrogen is, of course, highly explosive, but 'forming gas', made up of nitrogen and from 4 per cent to 8 per cent hydrogen, works almost as well and is much safer.) Once treated, this film can be kept at high sensitivity in its original plastic container for months in an ordinary freezer. When needed, the film is brought up to ambient temperature while still shut tight in its plastic can (I find an armpit to be a convenient warming chamber), and then used like any untreated film. Afterwards, the unused film can be re-frozen for later use still with little loss of sensitivity.

Cooking chambers and associated equipment can be purchased for not too much money and with full instructions from LUMICON, address given in Appendix J. Figure 7.3 shows their most popular model.

Colour, or black-and-white?

For discovery work, b&w has to be the first choice simply because urgency is usually required (at least, if you want to be the first) and home developing is quicker and easier. If you have never developed your own, you will find that the major photography firms have made it easy for you: kits are abundantly available with all the necessary equipment including convenient light-tight film tanks. No darkroom, other than a tiny closet for loading film into the tank, is needed. See the section on processing for additional remarks.

And so which b&w film is the best? Improvements are always being made, for the last couple of decades, one emulsion stands well above all the rest: Kodak's Technical

Pan (TP, formerly number 2415 or 4415). No other reasonable film has come close to having the extremely fine grain, high contrast, and high sensitivity at H-alpha (see below) of this astronomer's dream film. However, it almost has to be hypersensitised; directly out of the box, its speed is a patience-trying ISO 50. The cooking-in-hydrogen recipe described above can bring it up to about 400.

If Kodak TP is not available where you live, it can be ordered from LUMICON and from the major photography mail order houses. And if all else fails, use T-Max in either the 100 or 400 ISO version.

Colour films do have several compelling advantages: they are a delight to look at, and chromes, at least, give a positive image. Knowing the colour of your discovery can sometimes be useful in interpreting what you have found. And finally, because a colour emulsion is actually three emulsions (sensitive to blue, green, and red light) in one, the grain size is effectively reduced. From my own experience, I have found Fujichrome and Ektachrome 400 to have appreciably finer grain than Tri-X or even T-Max.

Much has been made of exposure-time guides for astronomy, but my suggestion is to stick to the good old reliable trial-and-error method. There are just too many variables like the brightness of your local sky and the transparency of the night sky. Note, however, that doubling the exposure makes a relatively small difference in an astronomical picture, especially b&w. If your wild guess is that an exposure ought to be somewhere around four or five minutes, start off by experimenting with exposures of one, four, and sixteen minutes. Afterwards, you can refine the time accordingly. In the discovery game, you will be taking many photographs, and soon you will become an expert at judging how to alter an exposure to take into account sky conditions, etc.

Limiting magnitude

It would be nice to have a chart that shows you how faint you can reach with different optical systems and different emulsions. But, as in visual observing, sky brightness and a murkiness of the atmosphere can be a cruel limiters, especially for diffuse, comet-like objects. As I stated in the previous chapter, at my home by the Pacific and 20 km from an urban area of a half million people, the faintest zenith star that I have ever been able to see is magnitude 5.4, just about a magnitude brighter than my limit at a really dark sky site. This same difference pretty much carries over to limiting magnitude on photographs. (But see section on filters below.)

The optimum exposure, the one that reaches faintest, will clearly show the presence of background sky light; on a b&w negative, this will mean somewhere between a medium and a dark grey. However, with super-fine-grained Kodak TP film, don't despair if the negative looks hopelessly black: with a sufficiently bright viewing light, stars near the ultimate limit can be picked out. With colour film, the ideally well-exposed sky will show up with some colour, and often not a colour that you would expect since the colour balance of the film may be strange at long exposure times. Fujichrome 400 has become the favourite of many astro-photographers partly because the sky colours come out an appealing rich dark blue.

As a general guide, you can count on reaching 11th magnitude with a well-exposed fine-grain film and a 50-mm lens even with a moderately bright sky. With 400 speed film at f/1.4, exposure times will be three to five minutes. The 200-mm diameter Schmidt described earlier reaches to 15th magnitude with similar exposure times, while longer focus telescopes with apertures this large and greater can record stars correspondingly fainter — but with considerably longer exposure times.

Again, as with visual observing, reaching the faintest magnitude involves picking out a faint image immersed in the sky background. For stars, generally speaking, the greater the focal ratio and the finer the emulsion grain, the fainter you can reach. More details will be found in Appendix D.

Along with the above, it should be remembered what was stated in the preceding chapter: It is one thing to pick out a star from the murk of the sky knowing that it is there; it is something else to find a newcomer for certain. Photographically, my experience has been to count on at least a full magnitude loss for discovery completeness. In other words, if you can detect stars to 12th magnitude, with care you should be able to discover novae, supernovae, or asteroids to 11th magnitude. And it would probably take a 9th or 10th magnitude comet to get discovered on the same photograph.

Filtering photographs

The filter principles outlined in Chapter 6 apply to photography: a sky lit up by moonlight or by garish bluish-green mercury street lights can be made darker with an orange or red filter. Similarly, light scattered from harshly yellow sodium lights can be suppressed with a red or with a blue-green filter. And comets, at least those that have wandered in fairly close to the Sun, can be enhanced relative to the background by special filters that transmit the colours of carbon emission. Consult the LUMICON catalogue (Appendix J).

Immediately after they erupt, novae and supernovae shine with a colour not very different from a star like Vega or Sirius, but within hours or a day or two, novae and Type II supernovae begin to glow especially brightly in certain wavelengths, owing to the presence of hot ionised gases such as hydrogen and nitrogen. These radiations, especially H-alpha in the medium deep red, are sometimes so strong, they can make the new star appear reddish to the eye. If a filter is used to isolate these emissions, almost any nova or supernova will stand out relative to most other stars.

To this information should be added that Kodak TP film was originally manufactured for solar physicists who were interested in photographing images of the Sun in the light of the red radiation of hydrogen, H-alpha. The reason is that Kodak TP emulsion has a sharp peak of sensitivity just at the colour of this emission, and when used with a suitable deep red filter, such as Kodak's Wratten Filter 29 or 92, or LUMICON's H-alpha filter, one has an effective nova/supernova detection system.

Furthermore, almost all distant stars in the Galaxy, including novae and supernovae, will have their light appreciably reddened by interstellar dust through which the light must pass. The red filter–TP combination will register all these far-off objects with high efficiency.

However, never forget that as you search for novae or supernovae, there is always the chance that a comet may come wandering into your field. Consequently, my own approach has been to compromise on the redness of my nova-search filter: mine is orange, specifically, a Hoya HMC O(G) filter.

Asteroids, shining by reflected sunlight, usually have colours similar to that of the Moon, and there is little that can be done to enhance their images. But, of course, almost any of the anti-sky or anti-streetlight filters will help.

Warning: a glass filter can cause internal reflections. Whenever one is used, be on the alert for false images (see below). There are two ways to avoid this potentially embarrassing situation: the first is to spend a little extra money and buy filters with anti-reflection coatings, such as the Hoya HMC series; the other is to use thin flexible gelatin filters like the Kodak Wratten series available at the better photography stores. Because they never lie perfectly flat, no sharply reflected images occur. However, these filters are easily scratched, and moisture can cause them to wrinkle and buckle. Because they are relatively cheap, buy some extras.

If the diameter of the objective lens cell is 72 mm or smaller, then glass filters can be purchased ready to screw over your camera lens. Gelatin filters will need a holder that you can easily make or can purchase. For Schmidts and telescopes, the filter will almost certainly have to be mounted directly in front of the film holder. Gelatin filters work well here since they are inside the telescope tube and protected from moisture; furthermore, they are so thin (usually 0.1 mm) that they do not alter the focal length appreciably as a glass filter would.

Most large mail order photography outlets can supply you with what you need, and of course LUMICON carries a variety of special filters useful for a variety of purposes.

Processing

Assuming that you will use the 'miracle' film, Kodak's Technical Pan, or Tri-X, then plan on using D-19 developer; and to maximise the amount of photographic information recorded in your exposures, develop for six or seven minutes (instead of the suggested four or five). T-Max should be used with its own special developer, or D-76. Also very important: agitate continually during development. To keep spurious images ('Kodak' stars or comets, as they have been cruelly and often unfairly called) to a minimum, gently rap the film tank on its bottom a couple of times (to remove bubbles) just after adding each new liquid, and use clean, particle-free water in the final baths. (I keep a cloth tea-filter handy for filtering doubtful water and solutions.) Then hang to dry in a dust-free environment. (A frequently cleaned tiled bathroom serves well.) The total processing time at 20 °C is about 18 minutes for fresh solutions, irrespective of drying.

As for colour film, the Kodak 'Hobby-Pac' gives you full instructions for processing any chrome film except Kodachrome. With it come the ingredients for the four needed solutions, sufficient to process six rolls of film (36-exposure, 35-mm). A precise temperature is not required, but knowledge of the temperature to within a degree or two is. Any temperature from 21 to 43 °C will do, but something around 38 °C (100 °F) is recommended. At this temperature, the total processing time, not including drying, is 30 minutes for fresh solutions.

After shooting film, my procedure is to retire to a small windowless bathroom in my home, lock the door, cover over the crack under the door with a towel, and unscrew the only light bulb over the sink. (The light switch is inconveniently – and dangerously – located outside the door.) The exposed film is then transferred from camera, or film box, to developing tank and the lid screwed on firmly. After that, I can come out into the light and start the developing process.

Inspection

Now comes the exciting part. To find a new comet, nova, supernova, or asteroid, the trick is to compare the most recent photograph – and always work with the original film or negative, never a print – with an older 'archival' photograph, preferably your own at the same scale and with similar exposure, sky brightness, and seeing conditions. You will want to get to the inspection as soon as possible: within hours, if possible, and certainly before the next night in case a confirmation photograph is needed. For comets, a careful inspection of your negative with a 15- or 25-power magnifier or microscope on a suitably illuminated viewing frame should be sufficient to turn up all fuzzy-appearing objects, and they can then be compared, one by one, with the archival negative. However, if your films reach much below 12th magnitude, you will start seeing dozens of faint galaxies and nebulae, many of them looking not unlike distant comets, and it then would be better to go to a *comparator*, a device that easily and quasi-automatically allows you to compare two negatives simultaneously or nearly so. For stars, you have no choice: you would be a straight-jacket case after comparing two deep negatives star by star; a comparator is a must.

In 1975, after serendipitously taking multiple photographs of Nova Cygni on the rise, Californian amateur Ben Mayer began to get interested in the possibility of making an inexpensive comparator that he and other amateurs could build and use to search for novae. He hit upon the idea of two slide projectors and a rotating shutter that would alternately flash first one, then the other sky negative on the same screen.

Fig. 7.4. A PROBLICOM, as used (and built) by Dan Kaiser.

(See Figure 7.4.) If one lines up all the stars and 'blinks' the two photographs, any newcomer to one slide signals its presence by blinking off and on. Ben calls it a PROBLICOM (PROjection BLInk COMparator).

The PROBLICOM idea works well, as many who have tried it can attest, but some of us, including Ben himself, have experimented with other blinking devices. As a result, several efficient methods have been developed, up to and including the use of a video camera to flash alternately one frame after the other on a TV screen — or subtract one frame from another leaving, in principle, only the intruders.

Howard Lazerson of Los Angeles decided that a better way would be to view the two negatives through identical magnifier–illumination systems and have the twin light sources flash off and on. It worked so well for him that he has marketed ready-made blink comparators. (See Appendix J for details.)

My own preference came about from a fascination as a boy with stereo viewers: my grandfather had one with a collection of faded photograph pairs of Egyptian pyramids, Paris nightclubs, and other sometimes naughty delights that I was not allowed to see. I am now using two 25-power 'macroscopes' purchased from Edmund Scientific (Appendix J) mounted at just my eye separation (65 mm) to look simultaneously at two 35-mm negatives. (See Figure 7.5.) Nothing blinks; I rely totally on my brain to tell me that something is amiss, that one eye is seeing

Fig. 7.5. The author's 'stereocom' using two 15 × magnifiers.

something that the other eye isn't. If the two photographs are taken about an hour or so apart, an image of a slowly moving comet or asteroid can appear stereoscopically out of the plane of the sky. For any type of variable star whose brightness has changed between exposures, the effect on the brain can be quite subtle, and it takes some experience to recognise the weak message relayed to the mind's interpretation centre. However, after blinking a few hundred frame pairs, I got the hang of it and can do as well as with a true blinker.

As in visual observing, the procedure is to scan the negatives slowly enough to check out all objects. And again, unless one is looking for variables, one should try always to concentrate on picking up faint objects; any bright comet, nova, super-nova, or asteroid will send a sufficiently strong signal to your brain.

A word of caution: spending more than an hour or so comparing photographs can be mind-blurring; it's best to interrupt searching at frequent intervals. I also find that listening to interesting (as opposed to soothing) music helps keep me alert.

Fig. 7.6. A comet trail? No, an emulsion defect sent to me by Manuel Lopez Alvarez of Buenos Aires.

Furthermore, it seems that one side of the brain processes sound and the other images; at least, I have improved my success rate appreciably ever since I started listening to J. S. Bach, Dire Straits, or Bill Evans.

Defects and false images

Can photographs lie? Unfortunately, yes, and in several ways. Inevitably, there will appear tiny dust particles, micro-bubbles in the emulsion, or mysterious specks, smudges, and splotches that sometimes can look arrestingly like a star or comet – like the discovery prospect shown in Figure 7.6. Do not be quick to criticise the film makers: considering the number of acres of high-quality film manufactured daily in various parts of the world, they do an amazingly good job. *Usually*, under higher magnification, defects show themselves as just that. Often a swish with a soft, clean, cosmetics make-up brush will dislodge your beautiful comet prospect.

The best prevention against mis-identification is a back-up photograph. Bountiful regions like Sagittarius should be photographed twice as a matter of routine, but for less productive areas of the sky, taking exposure pairs is not very cost–time effective, and you will just have to wait until the next opportunity to take a confirmation photograph, or call someone you know in the world where the sky is dark – or soon will be – for an independent observation or confirmation of a suspicious object.

Most insidious are the false images produced by internal reflections of bright stars, discussed already in this and the preceding chapter. Always be on guard lest there be a bright star in the field or, even worse, lurking invisibly just outside the field.

Search strategies

The vital questions of where and when to search have been outlined in earlier chapters and well summarised (except for asteroids) in the preceding chapter. For asteroids, you should review the pertinent part of Chapter 4 for the salient facts about their behaviour as seen from Earth. As Rob McNaught pointed out, most asteroids are best photographed near full phase, i.e., 180° from the Sun. And, of course, the Earth-threateners can appear anywhere in the sky and at any time which is, after all, what makes *any* astro-photograph exciting: one may turn up on the next exposure that you make.

To reach the faintest possible magnitudes, one should guide on the expected motion of the asteroids rather than on a star. For this you should have a means of moving the crosshairs by known amounts and directions with respect to the camera or telescope. Of course, if you are after main belt objects, and a bright one happens to be in the field, then guide on it.

It would also be well to review how the professionals use the 'magnificent discovery machine' of the Palomar Mountain Observatory, the 46-cm (18-inch) Schmidt telescope shown in Figures 5.7 and 5.19. In their contributions to Chapter 5, Eleanor Helin and Carolyn Shoemaker describe their similar but separate programmes with this f/2.0 super-camera. (Field about 9 degrees in diameter – 60 square degrees; focal scale of 225 arcsec/mm.)

But remember the Chapter 5 contribution of Brian Manning: telescope/cameras with a fraction of the aperture of the Palomar machine are discovering many of the new asteroids today. The impressive achievements of Japanese and Italian amateurs using telescopes that can and are easily built by ATMs (amateur telescope makers) or inexpensively purchased are especially noteworthy. Fundamentally, the reason this harvesting of asteroids is possible is the one already stated: professionals simply are not able to cover the entire sky on a continual basis, not even that 'hot spot' of main belt asteroids just to the east of the opposition point.

Final words

One strategy that should be obvious to all (but is often forgotten in the optimistic planning stages of new observing programmes) is to take photographs that can be used for the discovery of more than one type of object. For example, in a photographic comet/near-asteroid search, include as many galaxies taken from the

Thompson and Bryan 'Supernova Search Charts' as possible. If you then check for supernovae as you scan the films for comets and fast-moving asteroids, you will have significantly improved your chances of discovering something. And remember Planet X!

I end this chapter on a revealing topic: the success – and failure – of the PROBLICOM project. Ben Mayer's original project proposal, presented in several popular astronomy books and magazines and in illustrated talks given at astronomy meetings, received an enthusiastic response from several hundred amateurs from all over the world.

The disappointing part of the story is that of the hundreds of 'PROBLICOMERS', only a few of us, Dan Kaiser from Indiana, Mike Collins in England, and possibly two or three more, have been able to come up with discoveries. I can't believe that all those other blinkers around the world have not tried. And many, Ben included, have turned up variable stars and asteroids that were, sad to say, already known. What is the problem?

Dan has told his side of the story in Chapter 5. He was having no luck discovering novae, and so he began to track down the variable stars that he noticed varying from film to film. His success is impressive, especially considering the less than perfect skies of Indiana.

Mike Collins, using a 135-mm lens and 'Tech Pan' film, tells a similar story. He started looking for novae in October 1988 and now has a long list of variable star discoveries (over 20 at time of publication) – but no novae yet. The sky that Mike has to live with must surely be worse than Dan's.

My own record, some 16 novae, a comet, and around 30 as-yet-unreported variable stars in just over nine years, has also been achieved under less-than-ideal conditions: a large metropolitan area nearby and average weather conditions. For the Milky Way patrol, my Nikon first started out clamped to a $50 equatorial mounting with a jury-rigged drive motor and a 2-inch finder-scope; almost continuous guiding was essential. Later, I acquired a 200-mm Celestron Schmidt, and it and my trusty Nikon now ride together on a shiny motor-driven fork mounting manufactured by Celestron; only occasional guiding (with a 90-mm aperture Schmidt–Cassegrain telescope) is necessary now. The comet and three of the novae – in the Large Magellanic Cloud – were discovered on Schmidt photos; the other novae were found with either a 55-mm or an 85-mm f/1.4 lens.

What do Dan, Mike, a few others and I do that others seem not to do? Without any question, the answer is persistence, that 'magic ingredient' first mentioned in Chapter 1. Neither Dan nor Mike gave up when no novae came his way. And since I began my patrol in 1982, I have religiously tried to photograph the brighter parts of the Milky Way twice a week and inspect the pictures before the next night. Bad weather, bright moonlight, social commitments, and vacations all, from time to time, interfere with our routines, and annually the Sun gets in the way of the richest parts of the Milky Way. And of course occasional trips away from home interrupt the routine. But whenever possible, we stick to our schedule.

Minoru Honda's eighth nova discovery came just after I had begun patrolling, and for me it was a tremendous personal embarrassment. His latitude is 34° N and mine 33° S; the declination of the nova was − 26°! To make matters worse, when I looked at my most recent photograph of the region after receiving the news, there it was: I had just plain overlooked *an eighth magnitude nova*. Part of the reason was my inexperience at the time; it was only my 122nd exposure. (I took Exposure No. 3000 in July 1991.) But more importantly, that part of the negative happened to be richly endowed with dust specks; the nova was competing for attention with a host of several dozen unwanted imposters.

Consequently, I decided to clean up my act − literally − by filtering developing and rinsing solutions, keeping the work areas immaculate, and carefully brushing or blowing off negatives before looking at them. Three months later, I found my first nova and I was on my way.

What do we do with all the negatives that show nothing new? Keep them, and for several reasons. First, they give us a valuable collection of archival photographs with which to compare new negatives; when Dan or Mike finds a new variable star, he can go back and measure its light curve and, with luck, derive a period. Also, there is always the chance that something new will be on the film − something that you didn't find or weren't looking for. And, finally, there have been times when I was asked to measure the brightness of a variable star that was behaving strangely just at the time I happened to photograph it. Standard 35-mm slides take up little room (and they impress visitors mightily when they see the collection!). Most negatives are mounted in plastic (not cardboard) slide frames, and stored in boxes in a dust-free cabinet. But I will admit that many of my negatives − all the back-ups − are kept in numerical order, impaled in their sprocket holes on long nails, each labelled by region.

8

Going modern: electronic searches

If you had a million dollars to spend on the *ultimate* discovery machine, call it the UDM, what would it look like? First of all, it would depend on what you wanted to discover. Because comets, asteroids, and novae are going to be found in extended parts of the sky, a wide-angle system would be called for. And that especially applies to the search for Earth-approaching asteroids and variable stars which can turn up anywhere. Maybe something not unlike the 46-cm Schmidt camera on Palomar would work well. But let us come back to the question of optics later.

For extra-galactic supernovae, the telescope could be of intermediate size, i.e., perhaps larger than most amateur telescopes but smaller than most professional telescopes — perhaps somewhere around a half-metre aperture.

In any event, the UDM should be in a generally clear, dark-sky site at an altitude high enough to get above much of the atmospheric absorption, a particularly important consideration when working near the horizon. Again we will return to the question of site selection later.

Although not absolutely necessary, the motions of the telescope should be fully computerised so that it would set quickly and automatically on pre-determined and pre-recorded coordinates. Once there, the UDM would take a suitable exposure with an electronic imaging device, and then compare it, all automatically, with an archival image of the same field. If something new showed up, the position of the interloper would be checked with an up-to-date listing of recent discoveries, or with predicted positions of known comets and asteroids, again all automatically. If nothing was known near the interloper's measured postion, a telephone would ring in the astronomer's house perhaps many kilometres away, and an image of the new object would appear on a monitor screen, complete with information on coordinates, magnitude, size, and direction of motion (if any).

At this point, some 'smart' would probably be needed. Should the telescope's routine be interrupted for another image? Should another telescope, perhaps equipped with a spectrograph, be called into service to confirm the object's existence and analyse the light? To be sure, a message ready to be sent to the Central Bureau for Astronomical Telegrams might also be composed (automatically) by the UDM, but the astronomer would probably be needed for the final proof-reading.

How far are we away from having an UDM? Very close. In fact, several may already be here. Certainly, all the steps are now possible, and in fact nearly all are routinely in use today at the major observatories of the world. The Hubble Space Telescope and many ground-based telescopes can set to within a mere second of arc of a pre-programmed position. And many professional telescopes are used to record CCD images. These in turn can be stored in computer memories and manipulated in various ways, including subtracting one image from another.

Because the CCD is now used in many home video cameras, camcorders, and the like, electronic industries are highly motivated to improve this little wonder, and we are going to see many advances made in the next few years. Complete CCD systems for the amateur astronomer, ready to plug in and use, have recently become available, and by the time you read this, there undoubtedly will be many more on the market, each with something new and marvellous to offer – including a lower price tag. It will have to be left to the reader to stay abreast of new developments.

CCDs: charge-coupled devices

Through a microscope a CCD chip looks like a wall made of square bricks. Each 'brick' is, in reality, a light-sensitive cell, or picture element ('pixel'), and the entire chip may have anywhere from several dozen to a million or more tiny elements. In the CCDs currently used for optical work, a pixel size is no more than 0.05 mm on a side. At this writing, the very largest professional CCDs commercially available are about 2000 × 2000 pixels in size, and this translates to a chip measuring about 100 mm on a side – just under four inches. Any one pixel can be identified by its 'address' – its row and column number – and a properly programmed computer can sample any pixel by referring to its x- and y-positions.

When exposed to light, the pixels respond by giving off electrons at a rate proportional to the intensity of the incident light, and a computer continually tallies up the number of electrons generated by the pixels. In addition, there is 'noise', meaning that even in complete darkness a few electrons are released. This noise can make it troublesome to see faint objects, although usually more 'noise' is produced by the sky itself.

After the exposure has ended, the computer tallies up all the counts, keeping track of the pixel addresses, and then displays the entire frame, made up of the outputs of hundreds – or millions – of pixels, on a TV monitor. (See Figure 8.1.)

Comparing the frame with one made earlier and stored in the computer memory, can be done in two ways: first, the computer software may contain a blink program which displays the two frames exactly as a blink comparator does. The more sophisticated way is to subtract one image from the other. With either method, to compare the frame with another the computer has to make sure that the stars on the two images fall at the same xy-positions. If not, one frame has to be manipulated – rotated or translated – so that they do. The same has to be done with the average number of counts per pixel; they should be the same for a sample taken from the two frames. When all these refinements are made, the comparison can be made.

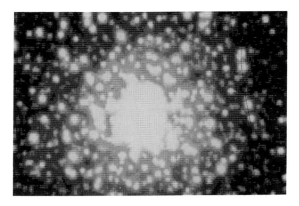

Fig. 8.1. The globular cluster M13 as imaged with a CCD and an f/14.6 11-inch telescope.

In principle it sounds easy, but in practice, variable seeing, variable atmospheric transparency, inaccurate tracking, differing background brightnesses, and maybe a half-dozen other effects can join forces to give system designers a headache.

Searching

One system already in use by the defence system of at least one nation (GEODSS of the USA) employs a number of telescopes around the world recording satellites and orbiting junk that pollutes the sky more and more each year. (See Figure 8.2.) Each telescope (1-m aperture, f/2) is programmed to take a series of quickly repeated short exposures, each long enough to reach a desired magnitude. For example, with one-second exposures they reach, under optimum conditions, 17th magnitude. Each exposure is recorded and stored – and subtracted from the first one. Because sky conditions and the system sensitivity change little between exposures, the quality of the image subtraction is virtually perfect and nothing is left to be seen – unless something has moved. Adding up five or six subtracted images will produce a neat little dotted line on the screen that shows the direction and speed of motion of the interloper.

With fast-slewing telescopes (15°/s) having a setting accuracy of a few seconds of arc, these 1-m telescopes can search 2400 square degrees per hour, or over 60 per cent of the entire accessible sky in one clear 10-hour night. An auxiliary Schmidt camera (15-inch f/2) can cover up to 15 000 square degrees per hour to a somewhat brighter magnitude limit.

If the time interval between exposures were increased to, say, ten or fifteen minutes, a relatively slow-moving object like a comet or asteroid would immediately become apparent. If the repetition frequency could be decreased to an image a day or a week, then the system becomes well suited for looking for variable stars, novae, and supernovae.

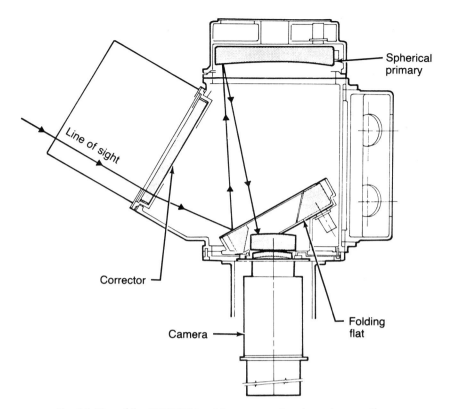

Fig. 8.2. One of the GEODSS 1-m telescopes used to detect faint satellites.

The University of California at Berkeley has developed an efficient supernova hunter: their UDM, maybe not quite ultimate but doing very well (in 1990), began by using a 0.9-m telescope in a poor site, and with this system they could easily reach to about 17th magnitude. (See Figure 8.3.) As Dr Carl Pennypacker, one of the chief scientists on the project, wrote to me, 'the realisation of our system has been made possible by recent advances in three areas: mini-computers, CCD imagers, and telescope automation systems. Shaft angle encoders, used in conjunction with personal computers for telescope control, presently have dropped in cost by a factor of one hundred since the 1960s when the pioneering work of Sterling Colgate was begun. Imagers have made tremendous advances: relatively low noise CCDs (5 electrons rms noise per pixel) have become readily available and fast workstations for real-time image processing are cheap and easily interfaced to.'

The observing procedure is much like that used by Reverend Evans except that a list of promising galaxies with coordinates is filed in the computer memory. Each waits its turn, and when it comes up to the meridian and the last exposure is completed, the telescope automatically slews over to the galaxy and starts the

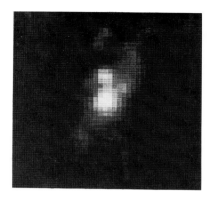 

Discovery image April 9, 1990   Reference image

Fig. 8.3. The supernova in NGC 3294 as discovered by the Berkeley Automatic Supernova Search, April 9, 1990.

exposure. However, in Pennypacker's words, 'the image processing has to be carefully coordinated with the telescope pointing, the file input/output, and the candidate handling. Generating search lists that are coordinated with a data base is an important feature of our system. We want to observe the prime galaxies at the correct time of the night, so there is an optimisation that must be performed on what gets observed when. All this takes a team of dedicated programmers and astronomers.'

Another approach used for finding new objects in the sky has been to have a dedicated telescope slewing back and forth along the meridian all night long. At its focus is a simple photometer, the output of which is fed into a computer memory. Each night's scans are compared with the previous night's. Again, if something moves or bursts into view, an alert message appears on the computer monitor with information about its position and brightness. If a meridional scan takes 20 seconds, then the aperture width (in right ascension) has to be at least 20 seconds, or up to five minutes of arc. Since some parts of the sky may be more interesting than others, then the scanning arc can be adjusted accordingly – to match the Milky Way or the ecliptic, for example. However, as we well know, comets, Earth-approaching asteroids, and the galaxies that might produce sufficiently bright supernovae can be found just about anywhere.

The beauty of this system is the simplicity of the telescope motions. The disadvantage, at least of the system just described, is that there is only one detector. If the telescope is set to scan over an 80° arc, then it will be slewing at a rate of 4° per second. If the aperture is five minutes of arc long, then that little piece of sky is sampled in 1/48th of a second. A 0.9-m telescope would be able to record an object not much fainter than 11th magnitude – and it would have to be in regions uncontaminated by brighter objects. Reducing the arc of the telescope slew to, say, 15° would improve the magnitude limit by almost two magnitudes.

But by using a large CCD, then an entire swath of the sky can be sampled. The University of Arizona's Spacewatch telescope, a 91-cm reflector on Kitt Peak, is programmed to scan the same region of the sky three times, once every half hour. Afterwards, the images are combined and searched by computer. If something moves (and is seen on all three images), the bells go off. One of their more spectacular discoveries was the Apollo asteroid 1991 BA which came within 170 000 kilometres of the Earth. (See Figure 1.2.)

The future for amateurs

Without question, these powerful systems, to say nothing of satellite telescopes such as IRAS, have a bright future. Does this mean a correspondingly grim future for amateur searches by visual or photographic means? Good Heavens, No! Not in the least. In the first place, there is the time element. Anyone can be the first to see something new in the sky, and it will be many years – if it ever happens – before the entire sky can be watched continuously by millions of patient CCD pixels. Also, never forget that the most powerful telescopes in the world still must be closed up on cloudy nights – or shut down because of technical reasons.

Late in the 19th century, when photography was just beginning to play an important role in astronomy, amateurs must have had similar worries. But just look at the discovery statistics. Today there are more visual discoveries of comets and extra-galactic supernovae than ever before, and amateurs still lead the field in photographic discoveries of novae and variable stars, and challenge the professionals in finding all but the faintest and fastest-moving new asteroids.

Amateur systems

The obvious way for amateurs to beat the professionals at their new electronic game is to join in and start using these marvellous new detectors themselves. As I write this, a number of commercially made CCD cameras are available at prices within the range of many amateurs. These packages also include some rather sophisticated image processing systems which allow the observer, using personal computers, to manipulate frames to match up stars and to blink one frame against another. And one does not need a crystal ball to predict that, within the next few years, many more systems will appear – and, it is hoped, at much reduced prices.

However, at the present time, the sizes of the CCD chip available to amateurs are too small in size to be used efficiently for anything other than searches in small areas – like supernovae in distant galaxies or variable stars in clusters. And so why not?

Similarly, telescopes that set automatically are here to stay and will quickly improve in quality – and decrease in price – in future years. Although none that I

Fig. 8.4. An image of the Ring Nebula captured by an SBIG CCD system and a 16-inch telescope.

know of yet exist for amateur telescopes, automatic positioners are becoming common, and all that is needed additionally are slewing motors and a bit more computer logic.

It would be foolish to use up any more space describing the details of existing systems, because they will soon be replaced by new and better ones. Suffice it to say that to make a go of it, new companies with a good product to sell will have to advertise. One company sent me a floppy disk which, when run in my PC, showed me exactly how the Ring Nebula would appear if I purchased both their hardware and software. (Their price was under $1000, not including the personal computer. See Figure 8.4.) The interested amateur need only read the advertisements and write away for information. Furthermore, journals like *Sky & Telescope*, *Astronomy*, and the *Journal* of the British Astronomical Association are continually reviewing and evaluating new products on the market. It was downright amusing to read (in the

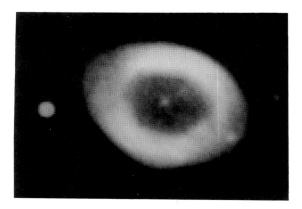

Fig. 8.5. A CCD image secured by Dennis di Cicco with an 11-inch telescope. This frame is actually the sum of four individual exposures.

September 1990 *Sky & Telescope*) the glowing comments made by my friend and veteran astro-photographer Dennis di Cicco. Even though he swears he will stick with photography – at least for a while – he was immensely impressed with how quickly and how easily frames reaching to the limit of his telescope could be acquired (see Figure 8.5) and then added or subtracted. I wonder how long he will stick with photography.

As they say, stay tuned . . .

9

Telling the world: reporting discoveries

Introduction

There are two kinds of astronomical discoveries that can be made: those for which there is little or no need to rush out and tell the world immediately; and those for which there is a definite urgency. In the first category one can include most variable stars – most pulsating stars, eclipsing binary systems, and dwarf novae – and most asteroids.

In Category No. 2 are those objects with transient characteristics, changing unpredictably and at times suddenly: comets, some eruptive variables (especially novae and supernovae), surface features or cloud patterns on the planets, and near-Earth asteroids.

Nevertheless, all newly discovered, run-of-the-mill variable stars should be reported reasonably promptly to the AAVSO or the BAA and later described in the IBVS, the Information Bulletin on Variable Stars. (See Figure 9.1.) New mainstream asteroids are the interest of the MPC, the Minor Planet Center; the Center will decide if your discovery is new or not, and, if so, calculate orbits and later bestow a number.

The CBAT, the Central Bureau for Astronomical Telegrams, is the internationally recognised clearing house for the urgent discoveries. Its parent organisation, the International Astronomical Union, meets every three years to review the policies of CBAT and, when needed, makes recommendations for changes.

For many years, virtually all urgent discoveries in the western hemisphere were reported to the Harvard College Observatory in Cambridge, Massachusetts, USA, from where 'Harvard Announcement Cards' – postcards – and telegrams were irregularly issued to subscribers. (For a while the author of this *Guide* served as editor.) In Europe, the information centre for such matters was the Central Bureau at the Copenhagen Observatory, and in fact the Central Bureau was established soon after the IAU came into being in 1919. When more rapid communication made two centres unnecessary, the Central Bureau was moved (in 1965) to the Smithsonian Astrophysical Observatory in Cambridge, Massachusetts, where it has remained ever since mainly under the capable directorship of Dr Brian G. Marsden (see Figure 9.2). Later in this chapter we will describe the operation of the Central Bureau.

The IAU also oversees both the MPC, again at the Smithsonian Astrophysical Observatory, and directed by Brian Marsden; and the IBVS – the Information

COMMISSION 27 OF THE I. A. U.

INFORMATION BULLETIN ON VARIABLE STARS

Number 3514

Konkoly Observatory
Budapest
5 September 1990

HU ISSN 0374 - 0676

PHOTOMETRY OF THE NEW ECLIPSING BINARY DHK 16 = SAO 80992

Kaiser (1990) discovered that the 9th-magnitude star SAO 80992 = BD +26°1996 is an eclipsing binary, which he designated DHK 16 in his discovery list. The spectral type is G0, the position RA 9^h 41^m 22^s, Dec +25° 35.1' (1950).

Regular visual monitoring by Baldwin soon detected additional minima. A discrete Fourier transform analysis of these visual observations by Kaiser produced several possible periods, the most promising being 0.69 day.

Williams observed the star photoelectrically with an Optec SSP-3 photometer and 28-cm Schmidt-Cassegrain. Most observations were made in the R band because of the variable's faintness with a 28-cm aperture and the photodiode's greater sensitivity at longer wavelengths. These observations (Figure 1) show that the 0.69-day DFT period is the half-period of an Algol-type eclipsing binary with nearly equal minima. A least-squares period solution using the discovery photo (HJD 2447968.691), one photoelectric minimum (the initial epoch of Equation 1), and six times of minima estimated from visual observations results in the preliminary light elements:

$$\text{Min. I} = \text{HJD } 2447999.617 + 1^d.3742 \text{ E} \qquad (1)$$
$$\underset{\pm.002}{} \quad \underset{\pm.0002}{}$$

The comparison star was SAO 80978 (7.26 V, +1.10 B-V, K1III). Williams measured the comparison star and the variable at maximum in the V and R bands relative to several nearby stars from the Arizona-Tonantzintla Catalogue (Iriarte et al. 1965) and obtained the following results:

Comparison = SAO 80978 = 7.26 V, +0.80 V-R
Var (max) = SAO 80992 = 9.22 V, +0.54 V-R

Fig. 9.1. A sample *Information Bulletin on Variable Stars*.

Fig. 9.2. Dr Brian G. Marsden.

Bulletin on Variable Stars – located at the Konkoly Observatory in Budapest. (See Appendix I for addresses and further information on these organisations.) The AAVSO and the BAA, both independent associations, have already been described (Chapter 3).

Making sure

Every discovery notice, whether urgent or not, must be accurate in all respects; *it is the responsibility of the discoverer to get the facts absolutely right*. A lot of effort and expense goes into the announcement of something new and potentially interesting, and if the facts are wrong, time and money, never in abundance at observatories, are lost – as will be the reputation of the one reporting the discovery.

Below you will find a list of suggestions for those just starting in the discovery game, but first you must commit to memory THE cardinal rule; it should be etched into your brain for all time:

BE ABSOLUTELY CERTAIN THAT WHAT YOU HAVE FOUND IS REAL

You must presume every candidate object guilty (of being a fake – a ghost image in the telescope, an emulsion flaw, a fly speck, a masquerading asteroid, or some other already known object) until proven innocent without the slightest penumbral shadow of a doubt. Amateurs (and this includes some professionals, too) should be deeply ashamed to know some of the sad statistics from the Central Bureau: Brian Marsden says about supernovae discoveries, '. . . we receive *two or three times as many erroneous reports* as we do of real SNe' [italics mine]; and CBAT Associate Director Daniel W. E. Green, writing about comet discoveries, states, '. . . for every real comet discovery, *there are perhaps four or five* that do not pan out' [again, italics mine]. *Four or five*. Incredible but true, and totally inexcusable.

Experienced observers eventually find out most of the ways that fakes can occur; the inexperienced have things to learn. Because the Central Bureau knows this well, beginners can expect to be thoroughly cross-examined, even after they have made their first several discoveries. So don't be annoyed when you try to report your first discovery; just make sure that your report is error-free.

Now for the suggestions, most of them obvious but all well worth repeating:

1. Since you will first check the identity of your find by comparing its position with positions of already known objects, supply yourself with a good set of star maps or galaxy charts (see Appendices G and H).

    Especially useful are precession tables, or a handy computer or calculator program that will convert coordinates. The reason is that some catalogues use a 1950 equinox; others 2000 or even 1900, or even earlier. You should have the means of converting rapidly and accurately from one coordinate system to another. Appendix B discusses precession in more detail.

2. Keep yourself up-to-date. If you do not live near an astronomical centre with a library, subscribe to journals covering your chosen field (see Appendix H) and to the appropriate IAU announcement service so that you will know about new discoveries. It also helps to know who made the discoveries, where, and how. Remember, too, that methods are always improving, and you will want to keep abreast of developments and new products.

3. Nova and supernova hunters should be particularly wary of asteroids and should have the means of finding out where the potential culprits are lurking – or count on waiting a sufficient period of time for possible movement to show up. Again see Appendix H for journals.

4. New discoverers would do well to have a friend or colleague check the object, either visually through a telescope or by inspecting critically the discovery negative. If you are a visual observer, it would be useful to enlist a colleague

located one or more time zones to the west, especially if you plan to observe shortly before dawn.

5. If you have a telescope that will provide higher resolution than your discovery instrument, use it to distinguish between a comet and a faint cluster, to look for motion, to note colour, and to discard possible ghost images. Also, change eyepieces and move the telescope around to rule out possible optical effects.

6. Photographers should have at least one, preferably two or more back-up photos . . . taken the night after, if necessary. Photographs *can* lie! The second exposure should be made with the camera pointed at somewhat different celestial coordinates so that the positions of any internal reflections from bright stars are grossly shifted. Photographers must also look at their photos promptly; it often does little good to report a possible discovery weeks after the fact.

7. An inexpensive 15 or 30° prism over the camera lens will produce a spectrogram adequate for distinguishing between a nova or supernova and an asteroid, or a long-period variable which brightened up just enough to be noticed. (See the end of Appendix C.)

Although there is often need for a super-rapid announcement − a nova or supernova on the rise, a comet disappearing into the twilight or moonlight, a fast-moving asteroid that could get away − haste must always be balanced against the limited patience of (1) the staff of the Central Bureau, and (2) the astronomers that may be asked by the Central Bureau to check on a reported discovery. Never forget the utter, utter shame of making an erroneous discovery.

Now, when you are absolutely sure

This section is concerned primarily with what to do when you really and truly have discovered something, and it should be announced to the world immediately. In other words, the following describes further the operation of the Central Bureau for Astronomical Telegrams and how and what you should communicate to it. Many of the same comments apply to discoveries of non-urgent phenomena that will be reported to other clearing houses.

News of discoveries comes into the Central Bureau by various means, but the two most reliable are telegrams (Telex or otherwise) or computer messages. In some locations, telegram offices are not always open; then you will have to telephone or relay the message to a friend in a big city where a telegram can be sent. Or, if all else fails, you can call the Central Bureau and deliver the message verbally. No matter what means you end up using, it is a good idea to send a confirming letter airmail; then there can be no doubts and the Central Bureau has an official record. (Postal addresses, TWX and telephone numbers, and computer addresses are given in Figure 9.3.)

For many years astronomers have used a standard code to send in reports of discoveries. While it is not necessary to put your message in code, you may wish to use it. It has been devised to minimise the chances for error in the transmission of the message; it guarantees that you have included all the necessary information, and it is concise. Full information about the code can be obtained by writing to CBAT. But, in any case, the following information should be included:

Your full name (and the names of other co-discoverers, if any).

Your address and telephone number (or other means by which the CBAT can reach you in case there are further questions).

The geographical location of the discovery site(s).

What the discovered object probably is (comet, nova, supernova, asteroid). Best to say 'Probable . . .' or 'Possible' comet, asteroid, nova, or supernova just in case.

The Universal Time/date of the observation(s); for example, July 4.1234, 1999 which would correspond to $2^h 57.7^m$ UT on the 4th of July, 1999. (Remember that there are 1440 minutes in a day; giving the date/time to a thousandth of a day implies an accuracy of 1.44 minutes, good for most discovery observations.) And you should state that you are using UT.

The right ascension and declination or, if you must, the offset coordinates from the centre of the parent galaxy, cluster, etc.

The equinox of the right ascension and declination (usually 1950.0 or 2000.0 AD).

The accuracy of the coordinates. If none are given, then it will be assumed that the final digit is close to correct. The same applies to the UT date/time.

Direction and rate of motion (if known and applicable). Position angles are measured from 0° due north through 90° due east, on around to 360°. The magnitude of the object (with clear information on any filters, and the detector used – eye, Tri-X Pan, Fujichrome 400, CCD).

The instrument used (e.g., 10 × 80 binoculars, 6-inch reflector, naked eye).

Description of the object (size of coma, length of tail, colour, etc.).

Comments on the object, for instance, possibly a re-occurrence of a nova outburst, close to but certainly not the same as a known comet or variable, reason for poor quality observation (e.g., very faint or very diffuse, cloud or twilight interfered). These comments are helpful to the Central Bureau if only to demonstrate that the discoverer is aware of other possibilities.

Additional observations, such as appearance with another telescope or on a back-up or confirmation photograph.

How discoveries are announced

When a discovery message comes into the Central Bureau, the staff first check to see if the report appears authentic and contains all the necessary information. If not, then

Circular No. 4814

Central Bureau for Astronomical Telegrams
INTERNATIONAL ASTRONOMICAL UNION

Postal Address: Central Bureau for Astronomical Telegrams
Smithsonian Astrophysical Observatory, Cambridge, MA 02138, U.S.A.
Telephone 617-495-7244/7440/7444 (for emergency use only)
TWX 710-320-6842 ASTROGRAM CAM   EASYLINK 62794505
MARSDEN or GREEN@CFA.BITNET   MARSDEN or GREEN@CFAPS2.SPAN

VW HYDRI

P. M. Kilmartin, Mt. John University Observatory, forwards the report by F. M. Bateson that this SU UMa-type dwarf nova is undergoing an unusually bright superoutburst, as indicated by the following visual magnitude estimates: July 12.35 UT, 8.5 (P. Williams, Sydney, N.S.W.); 12.69, 8.4 (A. Pearce, Perth, W. Australia); 15.25, 9.0 (A. Jones, Nelson, New Zealand); 15.71, 8.9 (O. Hull, Auckland, New Zealand); 16.15, 8.6 (M. D. Overbeek, Edenvale, R.S.A.); 16.27, 8.9 (Hull).

V854 CENTAURI = NSV 6708

W. A. Lawson, P. M. Kilmartin, and A. C. Gilmore, Mount John University Observatory, report a decline of this R CrB-type star, as indicated by the following photometry: June 22.52 UT, $V = 7.44$, $B-V = +0.63$; June 27.39, 7.44, +0.65; July 14.53, 8.04, +0.71; July 17.31, 8.18, +0.75.
  Visual magnitude estimates (cf. *IAUC* 4729): Jan. 25.75 UT, 8.0 (R. H. McNaught, Siding Spring Obs.); Feb. 11.57, 7.8 (P. Williams, Heathcote, N.S.W.); 24.46, 8.3 (Williams); 27.79, 8.2 (McNaught); 28.46, 8.4 (Williams); Mar. 2.47, 8.4 (Williams); 7.80, 7.9 (McNaught); 8.47, 7.8 (Williams); June 11.91, 7.6 (A. Pereira, Cabo da Roca, Portugal).

PERIODIC COMET BRORSEN-METCALF (1989o)

U. Fink, A. Schultz, and M. DiSanti, University of Arizona, report the following R (Kron-Cousins) magnitudes, obtained with the 1.55-m reflector (+ CCD) at Mt. Bigelow on July 13.46 UT: aperture 3″.8, $R = 17.1$; 9″.4, 15.7; 18″.8, 14.8; 37″.6, 13.9; 56″.4, 13.4. A pronounced ion tail was visible, as was a bright nucleus with a well-developed coma.
  Total visual magnitude estimates using 20×80 binoculars: July 14.44 UT, 8.1 (C. S. Morris, near Mt. Wilson, CA); 15.31, 8.2 (J. E. Bortle, Stormville, NY); 16.45, 8.2 (Morris, Pine Mountain Club, CA).

WX CETI

Visual magnitude estimates (cf. *IAUC* 4793): June 8.87 UT, 12.3 (Pearce); 12.71, 12.8 (McNaught); 16.75, 13.3 (Jones); 20.89, 13.3 (Pearce); 22.89, [13.9 (Pearce); 25.83, 15.6 (McNaught, photovisual); 27.83, 15.9 (McNaught, photovisual); July 5.78, 16.3 (McNaught, photovisual).

1989 July 18   *Daniel W. E. Green*

Fig. 9.3. A sample IAU *Circular*.

the sender must be reached and questioned. But if all seems well, then the report is coded and sent immediately to those who subscribe to CBAT's telegram service or electronic mail. Often confirmation or further observations will be requested of other observers.

Finally, the official notice, the International Astronomical Union *Circular*, or IAUC, is printed and mailed to subscribers. (See Figure 9.3.) These 9 × 13.5 cm cards contain not just discovery announcements, but ephemerides of comets and fast-moving asteroids, magnitudes of newly discovered eruptive or other interesting variable stars, reports on the appearance, spectrum, and behaviour of these objects, and announcements of world-wide observing programmes requested by astronomers manning large telescopes or satellite instruments.

As of September 1, 1990, monthly subscription cost US$10 (invoiced accounts), or US$6 (non-invoiced accounts). A version for electronic bulletin boards is also available. To give you an idea of about how many *Circulars* subscribers receive, in 1990 an average of about 18 were mailed each month: about three for a dollar.

If you are really planning to begin a regular search programme for comets, novae, supernovae, or Earth-approaching asteroids, or if you just want to be among the first to learn of the more sensational discoveries in the field of astronomy, then you should subscribe – or at least visit a library that does, or make friends with someone in an astronomical centre.

A few organisations have their own announcement services and usually provide additional information like extended ephemerides, finding charts, and additional reports by members, and not just on the newly discovered beastie but on other objects of special interest to the group.

Final words

I would, of course, be immensely pleased if I were to learn that the name of one of the readers of this *Guide* had just shown up for the first time on the cards of the IAUC – ACCOMPANIED BY ERROR-FREE REPORTS. And I am sure this feeling is shared as well by the majority of the contributors to Chapter 5.

There is something satisfying when an amateur, rank or otherwise, beats out a professional armed with all the optical (and electronic) might available. Please don't get me wrong: some of my best friends are professionals. I used to be one myself, too. And I certainly wish those who are involved in 'automatic' searches much success. But my heart leaps up higher when I behold a new 'amateur' Comet Bradfield than when I view a new 'professional' Comet Shoemaker. (Sorry, Carolyn.)

A good case can be made for the contention that astronomy is the only scientific field in which amateurs can make significant discoveries. Some would say 'can *still* make significant discoveries', but to me, that implies time is running out. I do not buy that argument. As long as there are cloudy nights on Earth, and as long as space astronomy remains relatively expensive (read here, 'not cost effective'), amateurs will have plenty of opportunities to make discoveries. All it takes are some optics, a little luck, and some of that good old-fashioned stick-to-itiveness.

And so I herewith end by wishing you good luck. Now, go get out the optics and start persevering.

Appendix A
Time

Not so many decades ago, the Sun and the stars used to be the fundamental markers of time. We still continue to maintain a system such that when the Sun crosses the local celestial north–south meridian, it is somewhere around what we call noon; likewise, when a point among the stars known as the vernal equinox (see below) is highest in the sky, it is 0 hours Local Sidereal Time.

Besides common, ordinary Standard Time, astronomical discoverers must deal with at least three other systems of time: Greenwich, or Universal Time, Ephemeris Time, and Sidereal Time. In this section of the appendix, each of these systems and a few others will be defined, discussed, and exemplified to the extent needed by most readers of this *Guide*. An excellent and more detailed treatment can be found in the *Observer's Handbook* of the Royal Astronomical Society of Canada. (See Appendix H.)

Seconds

Because of unpredictable quantities like the amount of annual snowfall the world over and how much sap runs up trees in the spring, the Earth has a slightly variable rotation rate. Days, hours, minutes, and seconds are now measured by cesium atomic clocks whose accuracy is equivalent to a second in about every 300 000 years. Currently, it takes, on the average, about 86 400.001 atomic seconds for the Earth to rotate once (a *mean solar day*), but to keep matters convenient, we make the day exactly 86 400 atomic seconds long, and correct the time by adding leap seconds on January 1 when necessary.

Standard Time

On the average, for every 15 degrees (°) of longitude eastward of any given point, the *time zone* is adjusted to an hour later. (In a few places like Central Australia, India, and Iran, half hours have been adopted.) But of course this concept of increasing time with east longitude cannot continue forever around the globe. Thus, in the middle of the Pacific is the *International Dateline* (IDL) where the date shifts to a day earlier *as one goes from west to east*. If you take the noon flight from Tonga to Samoa you arrive at 1 p.m. on the previous day.

Greenwich Mean Time

Because Greenwich, England, officially marks 0° longitude (see Appendix B), it is the logical place to standardise time. Therefore, many astronomical events are timed or predicted in one of the systems of Greenwich Time or, more specifically, *Greenwich Mean Time* (GMT), or *Greenwich Mean Solar Time* (GMST, not to be confused with *Greenwich Apparent Solar Time* which is awkward to use because the Sun appears to move at an uneven rate through the sky).

While in many parts of the Earth it is possible to get a reasonably accurate Standard Time signal from the radio or television, or by ringing up a telephone Time Service, short-wave radio provides the official, or Universal Time (see Figure A1).

Universal Time

Because of its universality, astronomers more frequently refer to GMT as *Universal Time*, or UT (sometimes abbreviated UT1), or *Coordinated Universal Time* (UTC). These times – UT, UT1, UTC, and GMT – are not all precisely equivalent, but for many of our purposes, they can be considered to be so. UT (often given as decimals of a day) is generally used for all phenomena involving the Sun or Moon such as eclipses and rise and set times; UTC is what one receives by short-wave. UT and UTC are never allowed to differ by more than 0.7 second.

Ephemeris Time

Because planets revolve around the Sun at highly uniform rates, it is convenient for people who calculate orbits to use a time system based on these motions. Therefore, in the ephemerides of comets and asteroids, positions are listed according to *Ephemeris Time* (ET). At one point ET was officially declared equal to UT, but over the years, because of slightly differing definitions, they have fallen out of step and now disagree by almost a minute. To correct matters, *Terrestrial Dynamic Time* (TDT), marked off by atomic seconds, has officially replaced ET, but ET is still used by most observer's handbooks and the Central Bureau for Astronomical Telegrams.

Sidereal Time

To set telescopes, one needs a time defined by the stars. The fundamental unit is the *sidereal second*, or 1/86 400th of the mean time interval between successive meridian crossings of the vernal equinox (VE or ♈), the point in the sky where the Sun crosses the celestial equator heading north. (For explanations of VE, the celestial equator, etc. see Appendix B.) By definition, the *Local Sidereal Time* (LST) is the time interval, in sidereal hours, minutes, and seconds, since the instant when the VE last crossed the celestial meridian of the observer. For no two observers will the LST be the same – unless they happen to be located precisely due north and south of each other.

Since the right ascension (RA or α) of a celestial body is the angle, expressed in

RADIO TIME SIGNALS

The list of radio signals below is thought to be of use to observers. The stations transmit Co-ordinated Universal Time (U.T.C.).The transmissions (except from OMA) also carry a coded correction so that it is possible to convert U.T.C. to U.T.$_1$: however, observers should always report their observations in U.T.C.

Station (Country)	Call Sign	Transmission Frequencies (kHz)	Times of Transmission	Details of Signal
Rugby (England)	MSF	60	00m 00s–59m 59s in each hour (except 1000–1400 on first Tuesday of each month)	second marker 100ms ⎫ interruption of minute marker 500ms ⎭ carrier wave Time codes for automatic equipment: (a) by lengthening or doubling of some second markers from 17s to 59s. and (b) within minute marker
Nauen (DDR)	Y3S	4525	continuous except from 0815–0945	second marker 100ms pulse minute marker 500ms pulse
Liblice (Czechoslovakia)	OMA (i)	2500	continuous except 0600–1200 on first Wednesday of each month	00m. 15m. 30m. 45m—call sign for 1 minute and 1 kHz tone for 4 minutes. For rest of period second marker 100ms pulse minute marker 500ms pulse
	(ii)	50	continuous except 0600–1200 on first Wednesday of each month	second marker 100ms ⎫ interruption of minute marker 500ms ⎭ carrier wave
New Delhi (India)	ATA	5000 10000 15000	1230–0330 continuous 0330–1230	second marker 20ms minute marker 100ms
Llandilo (Australia)	VNG	5000 10000 15000	continuous 2200–0700 2200–0700	second marker 50ms 55s–58s markers 5ms 59s omitted minute marker 500ms
Sanwa (Japan)	JJY	2500 5000 8000 10000 15000	continuous except 35m–39m	second marker 5ms pulse minute warning pulse 655ms from 59·045s–59·700s 9m–10m. 19m–20m. 29m–30m. 39m–40m. 49m–50m. 59m–60m call sign and voice announcement
Fort Collins (U.S.A.)	WWV	2500 5000 10000 15000 20000	continuous	second marker 5ms pulse (29s and 59s omitted) minute marker 800ms pulse male voice announcement 52s–60s
Kauai (Hawaii)	WWVH	2500 5000 10000 15000	continuous	second marker 5ms pulse (29s and 59s omitted) minute marker 800ms pulse female voice announcement 45s–52·5s
Ottawa (Canada)	CHU	3330 7335 14670	continuous	second marker 01s–28s 30s–50s minute marker 500ms pulse 51s–59s voice announcement, station and time. Long hour marker
Caracas (Venezuela)	YVTO	6100	continuous	second marker 100ms pulse 52s–57s voice announcement of time minute marker 500ms pulse 30s marker omitted

Fig. A1. Radio Time Signals (reproduced from the *Handbook* of the British Astronomical Association).

hours, minutes, and seconds, measured eastward along the celestial equator from the VE to the object, then it follows that the LST will be the same as the RA of all objects on the celestial meridian. For example, an object whose RA is 0^h 15^m will cross the celestial meridian fifteen minutes after the VE does, or at LST = 0015.

In practice, to calculate the exact LST, one needs a precise and current ephemeris such as the *Astronomical Almanac* or its somewhat abbreviated and immensely useful personal computer version, the 'Floppy Almanac' (see Appendix H), but many observer's handbooks give tables and formulae sufficiently accurate for most purposes.

For a rough rule-of-thumb method to estimate LST ('Star Time'), use the doggerel

> Each month has a number;
> Twice that gives you happily
> The 6 p.m. Star Time,
> Said Harlow Q. Shapley.

This rhyme time actually corresponds to the 21st of each month and, even so, is only good to an hour or so, but at least it tells you if the Orion Nebula (RA about 5^h) or central Sagittarius (RA about 18^h) will be good for viewing shortly after dark.

Time signals

In principle the atomic times kept by clocks in Greenwich, Washington, and Paris are averaged together, and this is what defines Coordinated Universal Time (UTC). UTC time signals are broadcast throughout the world by a number of short-wave stations whose frequencies and locations are listed in Figure A1, reprinted from the *Handbook* of the British Astronomical Assocation.

Appendix B
Coordinate systems

Circles and poles

Take a sphere, any sphere, slice it exactly in two, and you have defined a *great circle*; it is the rim of the hemisphere or, more specifically, that arc defined by the intersection of the sphere's surface and a plane passed through the centre of the sphere. It is the largest diameter circle that can be drawn on a sphere. Its *poles* are the two points that are 90 degrees (°) away from the great circle.

The *celestial sphere* is the entirety of the sky, infinite in radius, impossible to reach out and touch. Invisibly marked on it are several great circles frequently used in astronomy. One is the *horizon* which, because of atmospheric refraction, in reality lies about a half degree above the apparent horizon as seen from sea level; its poles are the *zenith* and the *nadir*.

The most important great circle in astronomy is the *celestial equator* whose poles are the two *celestial poles*. Their exact positions are defined by the terrestrial globe. The Earth's spin axis, extended to infinity, punctures the celestial sphere at the celestial poles; its equator, expanded to infinity, marks the location of the celestial equator. Because the most conspicuous apparent motion of just about everything in the sky is caused by the Earth's rotation, most things trace out circles that run almost perfectly parallel to the celestial equator. However, only objects precisely on the celestial equator move in great circles; all else moves in lesser circles.

Perpendicular to both the horizon and the celestial equator, there is a special great circle, the *celestial meridian*; it passes through the celestial poles, the zenith and the nadir, and the north and south points of the horizon. Any object crossing the meridian is said to be *culminating* or is *transiting the meridian. Upper culmination* occurs above the visible pole; *lower culmination* below.

Two other great circles of special interest to us are the *ecliptic* and the *galactic equator*. The first is traced out by the centre of the Sun in its annual path through the (usually) unseen stars. It is also marked by the intersection of the Earth's orbital plane with the celestial sphere. The galactic equator is that great circle which follows, as best it can, the central line of the Milky Way.

Coordinate systems

Each of the great circles has a starting point from which degrees are marked off. In the horizon system, we measure *azimuth* (*A*) starting at the north point and increasing through 90° due east, and on around. The angular height of an object above the horizon is the *altitude* (*a* or *h*) which will be negative for objects below the horizon. Thus, we have the *alt-azimuth* coordinate system, and at any instant, we can, in principle, measure these coordinates for any celestial object. The altitude of the visible celestial pole equals the latitude of the observer; its azimuth is either 0 or 180°, depending on the hemisphere in which the observer is located. The zenith and nadir are, or course, at plus and minus 90° altitudes; their azimuths are indeterminate.

Both the *equatorial* and the *ecliptic* systems have the same fundamental starting point: at one of the two intersections of the celestial equator and the ecliptic, specifically the point where the Sun crosses the equator heading northwards. Its name is the *vernal equinox* (VE or ♈), and the Sun is there on or about March 21 of each year. Ninety degrees away along the ecliptic (in both directions) are the two *solstices*; 180° away is the *autumnal equinox*. The *ecliptic latitude* (β) of an object is its angular distance north (+) or south (−) of the ecliptic, measured perpendicularly from the ecliptic; the *ecliptic longitude* (λ) of an object is the angle, measured eastward along the ecliptic, from the ♈ to the foot of this perpendicular. The latitude of the sun's centre is always 0°; its longitude increases at about 1° per day. The Moon, the major planets, and most minor planets all stay within a few degrees of the ecliptic, but because we see them from a moving observation platform, namely the Earth, the motions are neither steady nor constant. It took the genius of Copernicus to figure out just why this was so.

Declination (δ or simply Dec.) is exactly analogous to latitude: it is the angular distance north (+) or south (−) of the celestial equator, measured perpendicularly to the equator. *Right ascension* (α or simply RA) is the analogue to longitude: it is the angle measured eastward along the celestial equator from the ♈ to the foot of this perpendicular. Usually, but not always, RA is measured in hours, minutes, and seconds. Since 24 hours are equivalent to 360°, 1 hour = 15° at the equator (less elsewhere); similarly, 4 minutes equal 1°; 1 minute of RA = 15 minutes of arc, and 1 second of RA equals 15 seconds of arc. Beware of the confusion of two kinds of minutes and seconds! The *local hour angle* (t or LHA) is the angle, expressed in hours, minutes, and seconds, measured from the meridian westward (+) or sometimes eastward (−). (The hour angle of an object always increases with time.)

Every celestial object changes its equatorial coordinates slowly owing to the *precession of the equinoxes*; that motion caused by the twisting of the ecliptic system relative to the equatorial system resulting from the changing direction of the Earth's spin axis. This annoying nuisance will be further discussed later on.

As stated in Appendix A, the *local sidereal time* (LST) can be defined either as the

TABLE OF PRECESSION FOR ADVANCING 50 YEARS

If declination is positive, use inner R.A. scale; if declination is negative, use outer R.A. scale, and reverse the sign of the precession in declination

R.A. for Dec.−	R.A. for Dec.+	Prec. in Dec.	Precession in right ascension											Prec. in Dec.	R.A. for Dec.+	R.A. for Dec.−
h m	h m	'	δ = 85°	80°	75°	70°	60°	50°	40°	30°	20°	10°	0°	'	h m	h m
12 00	0 00	+16.7	+ 2.56	+2.56	+2.56	+2.56	+2.56	+2.56	+2.56	+2.56	+2.56	+2.56	+2.56	−16.7	12 00	24 00
12 30	0 30	+16.6	4.22	3.39	3.10	2.96	2.81	2.73	2.68	2.64	2.61	2.59	2.56	−16.6	11 30	23 30
13 00	1 00	+16.1	5.85	4.20	3.64	3.35	3.06	2.90	2.80	2.73	2.67	2.61	2.56	−16.1	11 00	23 00
13 30	1 30	+15.4	7.43	4.98	4.15	3.73	3.30	3.07	2.92	2.81	2.72	2.64	2.56	−15.4	10 30	22 30
14 00	2 00	+14.5	8.92	5.72	4.64	4.09	3.53	3.22	3.03	2.88	2.76	2.66	2.56	−14.5	10 00	22 00
14 30	2 30	+13.3	10.31	6.41	5.09	4.42	3.73	3.37	3.13	2.95	2.81	2.68	2.56	−13.3	9 30	21 30
15 00	3 00	+11.8	11.56	7.03	5.50	4.72	3.92	3.50	3.22	3.02	2.85	2.70	2.56	−11.8	9 00	21 00
15 30	3 30	+10.2	12.66	7.57	5.86	4.99	4.09	3.61	3.30	3.07	2.88	2.72	2.56	−10.2	8 30	20 30
16 00	4 00	+ 8.4	13.58	8.03	6.16	5.21	4.23	3.71	3.37	3.12	2.91	2.73	2.56	− 8.4	8 00	20 00
16 30	4 30	+ 6.4	14.32	8.40	6.40	5.39	4.34	3.79	3.42	3.15	2.94	2.74	2.56	− 6.4	7 30	19 30
17 00	5 00	+ 4.3	14.85	8.66	6.57	5.52	4.42	3.84	3.46	3.18	2.95	2.75	2.56	− 4.3	7 00	19 00
17 30	5 30	+ 2.2	15.18	8.82	6.68	5.59	4.47	3.88	3.49	3.20	2.96	2.76	2.56	− 2.2	6 30	18 30
18 00	6 00	0.0	15.29	8.88	6.72	5.62	4.49	3.89	3.50	3.20	2.97	2.76	2.56	0.0	6 00	18 00
0 00	12 00	−16.7	+ 2.56	2.56	2.56	2.56	2.56	2.56	2.56	2.56	2.56	2.56	2.56	+16.7	24 00	12 00
0 30	12 30	−16.6	+ 0.90	1.74	2.02	2.16	2.31	2.39	2.44	2.48	2.51	2.54	2.56	+16.6	23 30	11 30
1 00	13 00	−16.1	− 0.73	0.93	1.49	1.77	2.06	2.22	2.32	2.39	2.46	2.51	2.56	+16.1	23 00	11 00
1 30	13 30	−15.4	− 2.31	+0.14	0.97	1.39	1.82	2.05	2.20	2.31	2.41	2.49	2.56	+15.4	22 30	10 30
2 00	14 00	−14.5	− 3.80	−0.60	0.48	1.03	1.60	1.90	2.09	2.24	2.36	2.46	2.56	+14.5	22 00	10 00
2 30	14 30	−13.3	− 5.19	−1.28	+0.03	0.70	1.39	1.75	1.99	2.17	2.31	2.44	2.56	+13.3	21 30	9 30
3 00	15 00	−11.8	− 6.44	−1.90	−0.38	0.40	1.20	1.62	1.90	2.11	2.27	2.42	2.56	+11.8	21 00	9 00
3 30	15 30	−10.2	− 7.54	−2.45	−0.74	+0.13	1.03	1.51	1.82	2.05	2.24	2.41	2.56	+10.2	20 30	8 30
4 00	16 00	− 8.4	− 8.46	−2.91	−1.04	−0.09	0.89	1.41	1.75	2.00	2.21	2.39	2.56	+ 8.4	20 00	8 00
4 30	16 30	− 6.4	− 9.20	−3.27	−1.28	−0.39	0.78	1.33	1.70	1.97	2.19	2.38	2.56	+ 6.4	19 30	7 30
5 00	17 00	− 4.3	− 9.73	−3.54	−1.45	−0.43	0.70	1.28	1.66	1.94	2.17	2.37	2.56	+ 4.3	19 00	7 00
5 30	17 30	− 2.2	−10.06	−3.70	−1.56	−0.47	0.65	1.25	1.63	1.92	2.16	2.37	2.56	+ 2.2	18 30	6 30
6 00	18 00	0.0	−10.17	−3.75	−1.59	−0.50	0.63	1.23	1.63	1.92	2.16	2.36	2.56	0.0	18 00	6 00

Fig. B1. Table of Precession for Advancing 50 Years (reproduced from the *Observer's Handbook* of the Royal Astronomical Society of Canada).

local hour angle (LHA) of the vernal equinox or, equivalently, the RA of the celestial meridian.

Finally, astronomers interested in objects outside the solar system often refer to the galactic coordinates of an object. *Galactic latitude (b)* is the perpendicular distance away from the galactic equator, plus on the north side, minus on the south, while *galactic longitude (l)* is measured eastward along the galactic equator beginning at a point very close to the direction of the Galaxy's nucleus in Sagittarius. Any object seen in the band of the Milky Way will thus have a low galactic latitude – usually less than ± 10° or ± 15°. If there were no interstellar absorption, the richest parts of the Milky Way would be found near latitude 0° at longitude 0° where we are looking straight into the centre of the Galaxy, and at longitudes 90° (near α Cygni) and 270° (near λ Velorum) where we are looking longitudinally down the spiral arm in which the solar system is located.

Conversion from one system to another is a straightforward but somewhat messy calculation. However, calculator programs and computer software make the conversions simple. Some commercially available programs for personal computers are listed in Appendix H; those into programming can easily write their own.

Precession

The nuisance factor mentioned earlier, precession, can also be taken care of by calculator or computer, or a relatively simple table. (My favourite appears in the *Observer's Handbook* of the Royal Astronomical Society of Canada and is reproduced with kind permission of that Society as Figure B1.) Many such tables list the precession in RA and Dec. for 50-year intervals since most star atlases, maps, and charts give positions for the *equinox* of 2000, 1950, or 1900. Memorising a pair of limericks makes rough-and-ready 50-year conversions easy:

> In regions around the equator,
> The RA grows greater and greater.
> In minutes of time,
> Two-point-six does just fine,
> Unless you're a rule-of-thumb hater.

> At Zero and Twelve hours precisely,
> The change in the Dec. peaks up nicely.
> Sev'nteen, plus and minus,
> Arc minutes, your Highness,
> But at Eighteen and Six: zip, concisely.

In other words, at RAs of 6 and 18 hours, declinations stay unchanged while at intermediate right ascensions, one has to interpolate. At declinations numerically greater than about 40 or 50°, the RA corrections given by the above poem become unreliable.

Table B1. *Summary table of the most used astronomical coordinate systems*

System	Great Circle	Poles	Longitudinal	Latitudinal
Horizon	Horizon	Zenith, nadir	Azimuth (A) E from north point	Altitude (a or h) ($+$ and $-$)
Ecliptic	Ecliptic	Ecliptic poles (N and S)	Longitude (λ) E from vernal equinox	Latitude (β) ($+$ and $-$)
Equatorial	Celestial equator	Celestial poles (N and S)	Right ascension (α) E from vernal equinox or Local hour angle (t) W from celestial meridian	Declination (δ) ($+$ and $-$)
Galactic	Galactic equator	Galactic poles (N and S)	Longitude (l) E from galactic centre	Latitude (b) ($+$ and $-$)

The four most important coordinate systems and their definitions are summarised in Table B1.

Appendix C
Magnitude and colour systems

General concepts

Magnitudes and colours, although easily defined on paper, can be awkward and often difficult quantities to measure precisely. The most serious problems usually arise from the effects of the sky: its very presence is bad enough since, in reality, one always has to measure how much brighter a comet, star, or star-like object is than the sky in which it is immersed — and then compare it with other stars, galaxies, or nebulae which may or may not be nearby. The real problem is this: the sky is continually changing, both in brightness (even when there are no bright city lights or displays of aurora borealis or australis), and in transparency (both with respect to time and with respect to direction in the sky).

Then there are problems involved with the colour sensitivity of the detecting system, the stability and calibration of the detector, and, simply, the small amount of information that arrives on a star (or comet) beam from afar.

Originally the worst part of the problem arose from the charmingly casual manner in which ancient astronomers referred to star brightnesses: 'A star of the first magnitude' was the way of describing the brightest 20 or so stars in the sky. Hipparchus in Greece extended the rationale by dividing the naked-eye stars into five magnitudes.

At some point it was discovered that visually, brightnesses are sensed in a logarithmic manner. For example, consider a star as bright as a candle at 10 kilometres. Define its brightness to be 1 TKC (ten-km candle). The eye–brain combination will perceive a difference in brightness between this 1 TKC star and another star with a brightness of 2 TKC as equal to the difference in brightness between a 10 TKC star and a 20 TKC star.

When, in the 19th century, astronomers began to make quantitative measurements of star brightnesses, it was found that a typical 1st magnitude star was close to 100 times brighter than a 6th magnitude star. And the definition was adopted. If one works through the arithmetic, this statement is equivalent to saying that the difference in magnitude of two stars equals the logarithm of their brightness ratio times 2.5, or

$$m_1 - m_2 = 2.5 \log B_2/B_1$$

Table C1.

Magnitude difference	Brightness ratio
1	2.51
2	6.31
3	15.8
4	39.8
5	100.0
6	251.3
7	631.5
etc.	

where we use the logarithm to the base 10. In other words, a *difference* in magnitude corresponds to a *ratio* of brightness. These quantities are summarised in Table C1. (Most good photographic and CCD observers know Table C1 by heart since it makes judging exposure times simpler.)

The zero point of the magnitude system can be traced to an agreed-upon set of stars, but in practice, visual observers take what is given them – in the form of charts supplied by the AAVSO, the NZRAS, or the Thompson and Bryan *Supernova Search Charts*, or magnitudes measured photoelectrically for some reason or other. Incidently, the visual magnitude of that 1 TKC star given several paragraphs back is 3.5.

A physical property that can severely affect visual magnitude estimates is colour. Variable star observers know how tricky it is to determine the magnitude of a conspicuously red Mira variable by comparing it to normal white and yellow stars. Even worse, the colour response of people's eyes varies, not only from person to person, but also depending on the brightness of the source.

The normal eye under daylight conditions has peak response in the green part of the spectrum – at about 550 nanometres, close to the wavelengths of the green emission line of mercury (546.1 nm) and the oxygen radiation at 557.7 nm that makes many aurorae appear greenish. But under low light conditions when the retina's 'rods' take over from the 'cones', the maximum sensitivity shifts to about 500 nm, close to the wavelength of the strong blue-green oxygen radiation seen in planetary nebulae (500.7 nm). Figure C1 illustrates these changes graphically.

Modern practices

To measure precise magnitudes and colours of objects, professionals (and many amateurs) use photographic emulsions, or even better, photometers and CCDs, comparing the brightnesses as measured through different filters. The most popular

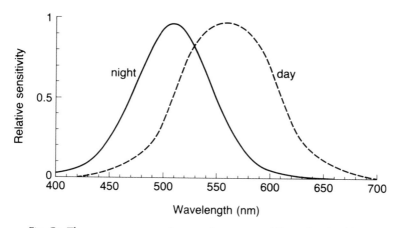

Fig. C1. The response curves of a normal eye at two different levels of light.

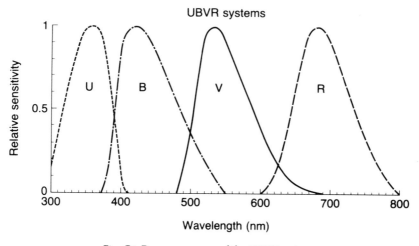

Fig. C2. Response curves of the *UBVR* system.

system in use today employs a combination of filters and detectors that produce, as well as possible, the response curves shown in Figure C2, marked *B* (for 'blue') and *V* (for 'visual'). Two other popular colours, *U* and *R*, are also shown. Notice that the *V* curve is not only narrower than the low-light-level response curve of the eye, but it peaks close to where the eye's daytime sensitivity is a maximum. We will return to this not-so-subtle difference shortly.

The *V* magnitude, when subtracted from the *B* magnitude, $B-V$, will obviously be a measure of colour: for example, a very red star at $V=10.0$ might have a *B*

Table C2.

Star name	V	$B-V$	Spectral type	Remarks
Achernar	0.46	-0.16	B3	Has weak emission lines
Aldebaran	0.85	$+1.54$	K5	Slightly variable
Rigel	0.12	-0.03	B8	
Capella	0.08	$+0.80$	G6 + G2	Close double
Betelgeuse	0.5	$+1.85$	M2	Variable
Canopus	-0.72	$+0.15$	A9	
Sirius	-1.46	0.00	A0	White dwarf companion
Procyon	0.38	0.42	F5	White dwarf companion
Alpha Cru	0.78	-0.25	B1	Close double
Beta Cen	0.6	-0.23	B1	Slightly variable
Arcturus	-0.04	$+1.23$	K2	
Alpha Cen A	-0.01	$+0.71$	G2	Alpha Cen B now 21 arcsecs away
Antares	0.9	$+1.83$	M1	Close double
Vega	0.03	0.00	A0	
Altair	0.77	0.22	A7	
Sun	-26.73	0.65	G2	

magnitude of 12.0, so that $B-V=2.0$. Red stars half hidden by interstellar dust clouds can have even larger values of $B-V$. The bluest stars in the sky sometimes have slightly negative $B-V$ values; the theoretical limit is close to -0.4, but for reasons that we will discuss below, that is not the limit in practice.

If we restrict ourselves to stars unaffected by the interstellar medium (the 'ISM') and for the moment forget about the influence of the Earth's atmosphere, we find a close correlation between colour and the kind of spectrum exhibited by the star. While we will not go into the details of spectral classification here, we will say that the spectrum of a bluish star ($B-V$ less than 0.2) is usually dominated by absorption lines of the elements hydrogen and helium, sometimes ionised. The spectrum of a middle-colour star like the Sun ($B-V=0.65$) contains in addition many metallic lines which dominate in the yellow stars where $B-V$ is around unity. By the time we reach $B-V$ greater than about 1.5, molecular absorptions take over and soon become so strong that they can obliterate large chunks of the spectrum altogether, especially in the blue.

Table C2 lists some familiar stars, all brighter than $V=+1.0$, and provides some sample colours and spectral types; many more can be found in most of the star catalogues listed in Appendix H.

The characteristic mainly responsible for the changes in spectrum and colour is temperature. On the surface of the hottest stars, temperatures can exceed 50 000 °C;

the Sun's surface is close to 6000 °C; and the red Miras have 'surface' temperatures of no more than two or three thousand degrees.

Emission lines

In a spectroscope, most solar system objects look very similar since they shine primarily by reflected sunlight, but of course comets near the Sun sport emission lines and bands. We will return to comets below.

Some of the most interesting stars, including novae and supernovae, will, like comets, often have strong emission lines throughout their spectra. These emission features can play tricks with the measured colours and magnitudes. For example, the brightest line of hydrogen emission, H-alpha, occurs in the red part of the spectrum (656.3 nm) and usually becomes strong in novae and Type II supernovae shortly after maximum brightness is reached. Because the wavelength of this line falls outside the response curve of the V system, the $B - V$ colour will be unaffected by its presence (although other lines of hydrogen will affect the B values and hence $B - V$). But the human eye responds, albeit weakly, to H-alpha, and a new nova or supernova, if bright enough, can appear reddish, even though the $B - V$ indicates the newcomer is blue or white.

Zero points

There exist many atlases and catalogues that give accurate magnitudes and colours of the brighter stars and nebulae. The most accessible and easily used of these are listed and briefly described in Appendix G. All stars down to about magnitude 10 have had their magnitudes measured respectably well, as have most galaxies and planetary nebulae to about 13th magnitude. Where needed, such as on the above-mentioned variable star and supernova charts, magnitudes of fainter stars are given. Finally, the individual stars in many clusters, open and globular, can be used to provide magnitude standards down to the faintest limits.

A particularly sticky problem arises when one wants to know reasonably accurately the magnitude of a newly discovered comet or star-like object that is fainter than magnitude 10 and that is located in the sky far from any handy magnitude reference. The best way I know of is to measure the diameters of star images on photographs, and one of the best, and accessible to amateurs, is the *True Visual Magnitude Photographic Star Atlas* ('TVMPSA') (see Appendix G for further information). All but the very faintest star images are impressively round and saturated, and their diameters correlate well with V magnitude. The formula I use with TVMPSA is

$$V = 13.9 - 11.6d$$

where d is the measured diameter in millimetres. At least in Sagittarius where the constants of this equation were determined, this relationship gives answers accurate, on the average, to about ± 0.2 of a magnitude. To reach fainter than magnitude 13.9, one must use one of the several deep-sky photographic atlases that can be found in the better-equipped astronomical libraries (or purchased for great quantities of money).

Comet magnitudes

As explained in Chapter 2, both predicting and measuring magnitudes of comets is something of an art. Estimating the magnitude of an ill-defined comet by comparing its total, or integrated, brightness with a set of stars of known magnitude takes much practice and experience. Visually, the usual procedure is to throw the telescope or binoculars out of focus so that everything appears more or less equally diffuse. Photographically, similar steps can be taken, but the resulting images are difficult to analyse at best. CCDs are best suited to the task since the sum brightness of all the pixels registering the comet can be directly compared with that of the comparison stars.

Predicting what will be the magnitude of a new comet would be easy if one knew exactly how the absolute brightness was going to increase as the comet neared the Sun. Almost always the brightness increases faster than inversely as the square of the distances to the Sun. Put into a formula, a comet's brightness, B, will vary inversely both as the nth power of its distance from the Sun, r, and as the square of its distance from the Earth, Δ, or

$$B = B_0 \times \Delta^{-2} \times r^{-n},$$

where r and Δ are expressed in AU, and B_0 is the comet's absolute brightness measured in some convenient units. If we convert this expression to logarithms and multiply through by -2.5, then each term becomes equivalent to a magnitude, and we end up with

$$m_1 = m_0 + 5 \log \Delta + 2.5\, n \log r,$$

where now m_0 is the absolute magnitude of the comet, namely, what its magnitude would be if it were 1 AU from both Sun and Earth. Values of m_0 can range greatly: Halley's Comet is around 5; Encke's Comet about 12. The subscript '1' in the equation identifies the magnitude as being that for the entire comet, as opposed to that for the nucleus alone where m_2 is used.

When a new comet is discovered, n is usually set equal to 4. An example: If the comet were reported as 12th magnitude (i.e., $m_1 = 12.0$) and then later calculated to be 3 AU from the Sun and 2 AU from the Earth, then the absolute magnitude

$$m_0 = 12.0 - 1.51 - 4.77 = 5.72.$$

As a comet nears the Sun, the n-value is re-evaluated and later ephemerides adjusted accordingly. Still, sudden changes in a comet's character, like losing its tail or having the surface of the nucleus crust over, can produce unexpected changes in brightness. The moral: be extremely wary of predicted comet magnitudes.

Asteroid magnitudes

Minor planets also can behave in an irregular manner owing to the nature of the hard, airless surface that we are looking at. Beside showing phases just like the Moon, an asteroid may have a relatively smooth surface, or one that is rocky and cratered. As described in Chapter 4, when an asteroid is seen exactly at opposition, all shadows disappear as seen from Earth. Consequently, there is a surge of brightness as the asteroid moves to within a few degrees of the opposition point. The same effect occurs with the Moon, of course, making some full Moons noticeably brighter than others since some can be over five degrees from the solar opposition point.

The formula used to predict the magnitude of an asteroid is much like that of comets. The differences are that n always equals 2 and a term gets added to take into account the phase. The formula is written

$$m(v) = H + 5 \log r\Delta - 2.5 \log F(a)$$

where $m(v)$ is the apparent visual magnitude, H the absolute visual magnitude, r and Δ the distances, expressed in AU, to the Sun and Earth, respectively, and $F(a)$ is a complicated correction that depends on the angle, a, between the asteroid and the Earth as seen from the Sun. This equation thus predicts magnitudes taking into account phase effects.

Finally, it should be noted that asteroids are not all alike. The easiest way to see this is to measure their colours accurately. To the eye the colours are all pretty much the same, but $B - V$ values show subtle differences that depend on the surface composition. The 'bluest' minor planets have $B - V$ colours a few hundredths of a magnitude bluer than the Sun, while the 'reddest' have $B - V$s close to $+1.0$.

Measuring colours

With a colour photograph, it is often possible to estimate the colour of a newly discovered object. The most accurate determination comes when the object is neither over- nor under-exposed. Of course, in principle, it should be possible, by comparing the object's apparent colour with nearby stars, galaxies, or nebulae of known colour, to estimate even its $B - V$ colour index, but in practice it is difficult because the degree of exposure will differ with each comparison object. In general, over-exposed objects appear too blue; under-exposed too red.

Paradoxically, black-and-white photographs and CCDs are better suited to the determination of colour indices than colour pictures. A pair of panchromatic photographs – taken with Kodak TP, for instance – one taken with a proper B filter, one with a proper V filter, can be used to determine B and V magnitudes for the newcomer. The difference of the magnitudes gives the desired colour index.

The problem for those who have only occasional need to determine $B - V$s is that the proper filters are expensive. However, one can do quite well by using a Kodak Wratten 47 filter for the B, and instead of V, measure a red magnitude, R, using a Wratten 25 filter.

Measuring magnitudes accurately from photographs is tricky, and details will not be discussed here. CCD determinations are more straightforward since one merely compares counts per star image (after having properly corrected the CCD frame for various instrumental effects). The point is, it can be done. Interested readers should refer to one of the more advanced texts listed in Appendix H.

Spectrograms

Another simple and very effective way of analysing the light of a newly discovered object is to put a prism or a diffraction grating in the light path. A 15 to 30° prism placed over the lens of an ordinary 35-mm camera converts the camera into a spectroscope with enough power to readily distinguish between a nova on the one hand, and an asteroid, a long-period variable, or an eclipsing binary on the other.

My procedure has been to use a prism (they collect light more efficiently and tidily than a ruled grating) in conjunction with a red filter and Kodak TP film. Here a telephoto lens is useful since it increases the dispersion of the colours. If the hydrogen-alpha line is present and bright in a star, the star's image will appear like a dot on a short line. The dot is the H-alpha; the line is the continuous light in the star's spectrum. The red filter is what reduces the length of the star's continuous spectrum to the short line. (See Figure C3.)

A small percentage of stars always have bright hygrogen lines in their spectrum, and so finding a star with bright H-alpha is not a guarantee that you have found a nova. Also, a small planetary nebula will appear as a dot, but usually with no line with it. (Again, see Figure C3.) A Type II supernova will have a broad H-alpha line — perhaps 15 nm wide — and if your spectrograph is powerful enough, it might show the line as slightly broadened.

Long-period variable stars stand out clearly since they always have strong molecular absorptions in their spectra. They show up on red spectral photographs as diffuse dashed lines. Their spectra come with various degrees of absorption, meaning that these dashes will vary in length from star to star and even with where in the cycle of light variation they happen to be. There are stars with moderately strong molecular absorption that do not vary, but they are rareties.

Photographically, comets are not the best of subjects unless they have a strongly condensed nucleus. Point sources are ideal for spectral studies. In principle, comets with bright and broad Swan bands (see Chapter 6) in the green and blue can easily be distinguished spectroscopically from galaxies (few if any emission lines) and

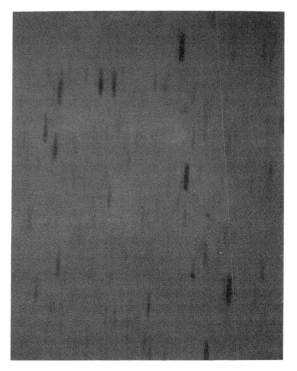

Fig. C3. The spectrum of the recurrent nova V 394 CrA (fuzzy dot near centre) when it was 9th magnitude. An 11th magnitude planetary nebula can be recognised as a fainter fuzzy dot above and to the right. Photographed by the author with a 15° prism, a 300-mm focal length camera, Kodak TP film, and a red (Wratten 25) filter.

planetary or diffuse nebulae (strong, sharp ionised oxygen lines at 500.7 and 495.9 nm plus hydrogen lines).

Visual observers should consider purchasing a spectroscope that attaches to the eyepiece of a standard telescope. If a nova or Type II supernova is bright enough (or the telescope big enough), the eye will be able to make out the second line of hydrogen, H-beta, which falls at 486 nm near the eye's greatest sensitivity. H-alpha at 656.3 nm may or may not be visible owing to the eye's low sensitivity in the red. But unless the nova, supernova, or comet is bright, seeing spectral features with the eye will be difficult at best. I encourage you to experiment.

Appendix D
Magnitude limits

Chapters 6 and 7 repeatedly refer to the goal of reaching to the magnitude limit and give some suggestions on how to get fainter. In this section of the Appendix we will look further into what is involved. To understand the problems can lead to making some progress in overcoming them.

General concepts

No matter what the detector – eye, photographic emulsion, CCD – the brightness limit to which one can ultimately reach depends principally on two factors: the amount of light reaching the detector and the amount of background light an image, stellar or otherwise, must overcome to make itself visible. Other factors can come into consideration, but they are minor compared to the above two.

Imagine the sky divided up into squares 1 second of arc on a side and containing faint star images 1 second of arc in diameter – i.e., good seeing. If the sky is sufficiently clear, the visual magnitude of one of these squares near the zenith will be around 21. If there happens to be a 21st magnitude star in one of these boxes, then the combined magnitude of the star plus bit of sky turns out to be 20.25. If the telescope plus detector – eyeball, photographic emulsion, or CCD – can reach that faint, then the star should be detectable.

It all comes down to whether or not the telescope–detector combination can capture enough photons of light during the effective exposure time to see differences between filled squares and empty squares of sky. In principle, it matters little whether you have a 10-m telescope or a 10-cm telescope; you just have to wait longer with a smaller aperture – unless the seeing is so good and the optics so good that the giant telescope perceives star images appreciably sharper than the modest instrument.

As a detector, the eye leaves much to be desired. Besides having a restricted wavelength range and a small aperture (see Chapter 6), its effective exposure time is of the order of a tenth of a second. However, as noted in Chapter 6, the brain's memory works together with the eye and can store visual information for many seconds, thereby making it possible to accumulate knowledge about the existence of a faint object in the field.

But note that 'seeing' is the Great Equaliser, especially when searching for faint stars. If unsteady atmospheric conditions explode star images to many seconds of arc in diameter, then the limit of a telescope depends almost wholly and exclusively on the number of photons captured, no matter whether by eye, photograph, or CCD. It may take a while, but in miserable seeing, a Nikon camera can come close to matching the magnitude limit of the Keck monster. However, the poor visual observer, hampered by a short-exposure detector – the eyeball – still needs a big telescope to collect a sufficient number of photons to register faint objects.

Unfortunately, nearly all of us have to work with skies lit up by nearby cities and towns, and few of us have access to telescopes of really large aperture. But the bottom-line problem remains basically the same: the challenge is to pick out the image, whether it be that of a star, comet, or asteroid, from the sky background. One extreme example is finding Venus with the naked eye in the daytime sky: it can be easily done if the sky is a dark, rich blue and Venus is near maximum brightness, but a little haze and a somewhat pale Venus make it wellnigh impossible. The resolution of the naked eye is close to a minute of arc, and so we must think in terms of squares of sky about one minute of arc in size, but the arguments are exactly the same: if the square containing Venus is imperceptibly brighter than adjoining squares, then Venus will be invisible.

Chapters 6, 7, and 8 list many additional factors that come into play when trying to reach to the limit of the eye, the photograph, and the CCD, but most reduce to bettering or somehow altering the optics, the observing site, and the observer's experience. Therefore, the graph that appears as Figure D1 should be used more to provide a *relative* indication of what bigger optics and differing magnifications can ultimately do. Unless otherwise stated, all indicated values correspond to reasonably good observing conditions – near the zenith where a 6th magnitude star is visible.

Visual observations

In a thorough study, Bradley Schaeffer presented an analysis of all the factors involved in seeing to the limit and compared the theoretical predictions with actual observations solicited from observers. Chapter 6 discusses the individual effects, while Schaeffer's results are summarised nicely by Figure D1.

In connection with this figure, make note of the following: The eye's pupil must be larger than the beam of light coming from the eyepiece. Using binoculars increases the indicated limiting magnitude by about 0.4 magnitude. With increasing age, the pupils of fully dark-adapted eyes open less wide; furthermore, the eye lenses yellow and lose some of their transparency. Having one's natural lens replaced with a plastic implant can do wonders. (I speak from happy experience.)

Photographic observations

For a given telescope aperture, the two most important factors involved in setting the magnitude limit on a photograph are the focal scale, usually expressed in minutes or seconds of arc per millimetre, and the emulsion characteristics. Considerations of

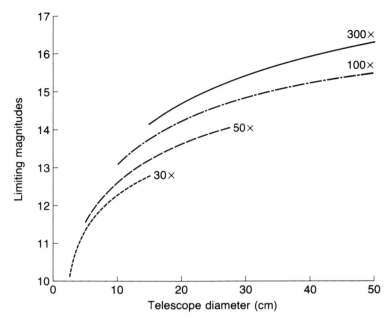

Fig. D1. The limiting magnitude of visual telescopes of various sizes and magnifications. (Adapted from Bradley Schaeffer's article in the *Publications of the Astronomical Society of the Pacific*, Vol. 102, p. 212, 1990.)

filter factors, developing times, and other influences have been discussed in Chapter 7.

As was noted in that chapter, under a high-powered microscope one can see that a photographic emulsion is composed of millions of individual grains. These are discrete particles of metallic silver embedded in the gelatin base. With lesser power one sees these grains as clumps, and these are what set the limiting resolution of the emulsion (see Figure 7.2). Remember that, in general, the faster the emulsion speed, the larger the clumps.

How limiting magnitude depends on this property of photographic emulsion is somewhat complicated but very much worth understanding. Here follows a résumé of what must be considered.

The focal scale and the size of the individual emulsion clumps should be considered together: for optimum performance – to have the highest emulsion speed commensurate with finest resolution – the size of star images, as smeared by seeing, should be matched to the linear size of the emulsion grain clumps. For example, consider an f/5 200-mm (8-inch) telescope. Its focal length, F, is $5 \times 200 = 1000$ mm, and the scale at the focus is

$$\text{Focal scale} = 206\,265/F = 206 \text{ arcsecs/mm.}$$

If the seeing were 2 seconds of arc, then the diameter of star images would be close to a hundredth of a millimetre, or 10 micrometres (μm, microns, or thousandths of a millimetre). Grain clumps in a moderate speed film – around ISO 100 – are just about equal to this size. A longer focal length telescope could use faster film, but for cameras of lesser focal lengths, in the range of 25 to 250 mm, one must use the extremely fine grain available in Kodak's Technical Pan (TP) film to reach the resolution limit.

Exposure time also depends on focal ratio, of course, and thus one wants to use an f-ratio that keeps exposure times reasonable – as well as providing acceptable resolution. Again there is a trade-off: for a given aperture, smaller focal ratios mean shorter focal lengths and reduced resolution.

Returning now to first principles, remember that the ultimate photographic limiting magnitude depends only on being able to pick out star images from the sky background. The problem boils down to having enough photons captured per star image, and it all becomes a matter of statistics: if a 1-second-of-arc star image has been produced by 11 photons, and surrounding 1-second-of-arc pieces of starless sky have, on the average, been exposed to ten photons each, the detection of the star is difficult if not impossible. A 0.1 magnitude difference is just at the limit of detectability.

Because calculating the limiting magnitude of a certain lens or telescope when used with a particular emulsion depends on many factors, (some rather ill-defined like 'seeing', grain size, and sky brightness), one is best off going by experience. Read here, 'Trial and Error'. Try various films with various exposure times and see which combination reaches faintest. Of course, if possible, match grain clump size to image size, but with a short-focal-length camera, then use the finest acceptable grain clump size.

A few words about emulsion speed. First of all, it should be remembered that most advertised film speeds apply only to short exposures. In general, at exposure times longer than a second or two, the stated ISO ratings drop, sometimes precipitously. This effect, called 'reciprocity failure', is especially severe for fast, coarse-grain film; it means that doubling the film speed does not compensate (reciprocate) halving the exposure time.

Chapter 7 described the two most popular methods to reduce this unhappy characteristic: to refrigerate the emulsion while exposing ('cold cameras') and by hypersensitisation of slow, fine-grain film (which usually suffers less reciprocity failure).

The so-called 'miracle film', Kodak's Technical Pan, received this immodest nickname because of its super-fine grain and high contrast – and its ISO rating can be increased by hypersensitising from around 50 to as high as 400, thereby making its fine grain available for almost any kind of astronomical work.

Because they are made of three-layered emulsions, colour films effectively produce finer grain than an equivalent speed black-and-white film. However, colour

Table D1. *Some popular emulsions and their characteristics*

Emulsion type	ISO*	Effective grain size (μm)**	Contrast
Kodak Tri-X 400	400	20	Medium
Kodak T-Max 400	400	10	High
Kodak Technical Pan (TP)	40	2	Very high
Kodak 103a-***	320	20	Medium
Kodak IIIa-***	35	8	High
Fujichrome 400	400	8	Medium

Notes:
* ISO film speeds (or ASA) can be increased by hypersensitising. However, in this table only Technical Pan and IIIa- show substantial gains.
** Grain size is usually indicated by the number of lines per millimetre that can be resolved. Point sources are not so easily resolved, and the values listed give the approximate diameter that a point source would appear to have.
*** The spectroscopic emulsions like 103a- and IIIa- are available with a variety of spectral sensitivities. For further information, see Kodak Publication P-315 and pamphlets on individual emulsions.

film does not hypersensitise as well as black and white because of the increased emulsion thickness.

In Table D1 I list several emulsions and their characteristics. Some data are taken from Kodak publications; some come from my own experience.

Users of telescopes or cameras with focal lengths greater than about 300 mm should consider carefully grain size in conjunction with the formula above giving focal scales as a function of focal length. The procedure is first to calculate the number of micrometres that are equivalent to a second of arc for your telescope; see the example above and Table D2. (With exposures lasting more than a few seconds, you can expect star images to be close to 1 second of arc in diameter in good seeing.) Then select the appropriate emulsion type whose grain clump size matches the seeing. But as stated earlier, for shorter focal lengths, you must decide on the best compromise between speed and resolution.

Table D2 lists a number of camera focal lengths and gives working values for focal scale and frame size.

CCD observations

Many of the considerations above are directly applicable to CCDs. However, instead of having grain clumps of irregular shape and size, CCDs are made up of square

Table D2. *Some standard camera focal lengths and what they yield*

Focal length (mm)	Field of view (°)*	Focal scale (arc min/mm (μm/arcsec))
28	49 × 74	122.8
35	39 × 59	98.2
50	27 × 41	68.7
85	16 × 24	40.4
105	13 × 20	32.7
135	10.2 × 15.3	25.5
200	6.9 × 10.3	17.19 (1.0)
300	4.6 × 6.9	11.46 (1.5)
600	2.3 × 3.4	5.73 (2.9)

Note:
* Field size as it would appear in a standard 35-mm film frame.

pixels with precisely known dimensions. The speeds of these individual pixels are much greater than the photographic grain clumps. But CCDs have decided disadvantages: the largest (and very expensive) CCD chips are only a few centimetres on a side; pixel sizes are large compared to grain clumps; and the cost and complexity of a complete and sophisticated CCD operating system is a major consideration, if not staggering, to most of us.

To achieve good resolution, professionals must use CCDs at large focal scales in order to have pixel dimensions of a second of arc or smaller. Fortunately, CCDs are about 50 times faster than the fastest film available, and consideration of f-ratios is not so important. Furthermore, reciprocity failure is non-existent at low light levels, although saturation of pixels by bright stars can be troublesome, especially if there is 'bleeding' of the signal to adjacent pixels.

Most currently available CCDs have good sensitivity from the deep violet to the near infrared with greatest response coming in the orange or red parts of the spectrum. Because they must be operated at cold temperatures, chips must be protected from dewing over in humid air by a window, usually warmed, which can, in fact, be a coloured filter.

To reach the limiting magnitude with a CCD is to perform a statistical analysis on the data. Elaborate computer programs exist which fit every possible star image to a standard and expected size and shape, add up the number of photons counted inside the boundary, and then compare the sum total to that produced by the surrounding sky background. If the difference star-minus-sky is statistically significant, the image is proclaimed real. Such sophisticated procedures have made it possible for professional telescopes to reach regularly to 25th magnitude and even fainter.

In most modest CCD set-ups, pixel sizes will probably be larger than star images, statistical analyses are largely ignored, and the final frames are inspected much as would be photographic negatives displayed on a television screen. As a result, the limiting magnitude will be very comparable to what one could achieve photographically. Perhaps the most used CCD reduction program would be one to subtract frames, one from another – to find what is left over. While a similar operation can be made by superimposing a photographic negative over a positive, preparing the positive means an added and exacting step in the darkroom – and, for various reasons, the result is not always satisfactory.

As Chapter 8 indicates, astronomical discoveries are now being made routinely with telescopes equipped with sensitive CCDs. All that is needed is sufficient computer power and a substantial outlay of money. The latter requirement limits the number of observatories that can afford to establish CCD search programmes.

Appendix E
Orbits and orbital elements

General concepts

The two-body problem: the study of the motions of two masses subjected to their own mutual gravity. Except on paper, there is no such thing of course. Every object in the universe attracts every other object in the universe. However, in the solar system, the Sun dominates, and when a new comet or asteroid is discovered, the initial orbit calculations can be made with sufficient accuracy assuming no other objects exist in the solar system.

The trick is to know how to take a series of observations of right ascensions and declinations and come up with an orbit – and then calculate future positions as seen from our spinning, orbiting, precessing, nutating platform, the Earth. The methods, although perfectly straightforward, are too complicated to describe here; books that tell you how are listed in Appendix H. However, graphical solutions that can give answers with an accuracy of five or ten degrees are possible, and they will be described below.

As Newton first showed, a comet or asteroid can move in one of four possible orbits: circular, elliptical, parabolic, and hyperbolic. In the solar system, only elliptical orbits persist: the chances of an object following a *perfectly* circular orbit are infinitesimal; and anything in a parabolic or hyperbolic orbit will soon be lost and gone forever. However, since most of the major planets and many asteroids move in *almost* circular orbits, and since many newly discovered comets move in *almost* parabolic orbits, we will find it convenient to look carefully at these two shapes, both of which are extreme examples of ellipses.

Hyperbolas should not be discarded completely: an object can, on occasion, have a close encounter with a major planet and have its orbit gravitationally altered. As a result, it is possible for a nominally stable member of the solar system to be sent off in a hyperbolic orbit never to return again. More interesting are comets that arrive from far outer space already in a hyperbolic orbit. Comet Furuyama 1988 IV was one: so was Černis 1983 XII. Both comets came in and went out well above the orbits of the major planets; therefore, the orbits were left pretty much unaltered as they passed through the solar system. It sort of suggests that these comets didn't just *fall* towards the Sun (in an elliptical or parabolic orbit), but were *accelerated* towards the

Sun by something far away, like another star or a comet that passed by, way out in the Oort Cloud (see Chapter 2).

These 'hyperbolic comets' deserve more study because they may bring us valuable clues about conditions in outer space. But here we will add no more except to say that their orbits are so close to being parabolic, we can treat them as such.

In any two-body problem we should, of course, really think in terms of the two objects revolving around a common point, the centre of gravity, but, in the solar system, this mathematical point will always be near the Sun's centre, especially so for comets and asteroids. Therefore, hereafter we will assume the two, the centre of gravity and the Sun's centre, to be one and the same.

Circular orbits

A circular orbit of a planet about the Sun can be located uniquely in space with just three *orbital elements*: the *radius*, R; the tilt of the orbit plane relative to that of the Earth, or *inclination*, i; and the orientation of the line of intersection of these two planes. Since this line of orientation must, by definition, lie in the plane of the Earth's orbit, it will, when extended out to the celestial sphere, intersect the ecliptic (see Appendix B). The orientation of the line of intersection is usually given as the *longitude of the ascending node*, the symbol for which is Ω. (Astronomically, a 'node' is one of the two points where the planet's orbit passes through the Earth's orbital plane. By 'ascending', it is meant coming up from the south to the north side of the ecliptic plane.) Ω, then, is the celestial longitude of the ascending node *as seen from the Sun*. As one might expect, the *descending node* lies on the same straight line passing through Ω and the Sun.

If the direction of motion of the planet is like that of the Earth – i.e., counterclockwise as seen from the north – then the inclination i is defined as less than 90°. But if the motion is in the opposite sense, then i is between 90 and 180°. These directions of motion are called *direct* and *retrograde*, respectively.

To specify where the planet is in this orbit, a further quantity is needed – like at what instant the planet is at the ascending node.

The planet's orbital period P is easily calculated by one of Kepler's laws: $P^2 = R^3$ where R is in AU and P is in years. From this, and with a little algebra, one can show that the planet's velocity, V, is given by

$$V = \frac{29.8}{\sqrt{R}}$$

(1)

where V is in km/s and R is again in AU.

Elliptical orbits

First, what is the exact shape of an elliptical orbit? It is the path followed by a small mass in the solar system such that the distance from the Sun *plus* the distance to some

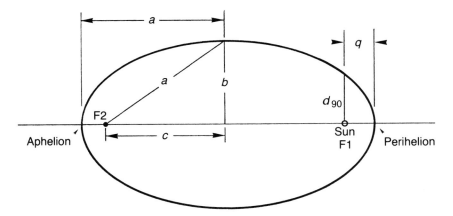

Fig. E1. An ellipse. The basic formula is $x^2/a^2 + y^2/b^2 = 1$. From this equation, the eccentricity, $e = \sqrt{(a^2 - b^2)}/a$.

other imaginary point F_2 is always a constant. This ellipse will have two *focal points*: the Sun is located at one *focus*; the other is at F_2. The distance from either focus to the centre of the ellipse is denoted by c, and half the length of the ellipse's long axis, the *semi-major axis*, is denoted by a. Figure E1 shows an elliptical orbit with these several quantities.

To locate an elliptical orbit in space, we need a total of five orbital elements. The first of these is a, twice which gives us the overall length of the orbit. The shape of the ellipse is given by the *eccentricity*, e, defined by c/a. So the rounder the orbit, the closer F_1 and F_2 become. For a circular orbit, the two focal points become one, both c and (therefore) e will be zero, and a becomes the radius. As the focal points separate, both c and e increase, and the ellipse begins to stretch out until c is so large, it approaches the value of a, and e approaches the value of unity. At infinity, c can become equal to a, and we then have a parabolic orbit; and since c now equals a, $e = 1$. More on parabolas in the next section.

In the solar system the Sun will always be at one focus (nothing significant will be at the other), and the minimum distance between Sun and planet (or comet) occurs when the planet is at *perihelion*. The *perihelion distance*, q, is given by $q = a(1 - e)$. In other words, q is the distance to the focus measured along the axis of the ellipse.

The semi-major axis, a, and the eccentricity, e, together with the inclination, i, and the longitude of ascending node, Ω (see preceding section), give us four of the necessary orbital elements. We need one more, the *argument of the perihelion*, called ω (lower case omega). It is defined as the angle, *as seen from the Sun*, between the direction to the ascending node and the planet's perihelion point, and is always measured in the direction of the planet's motion.

But something is still missing: to indicate *where* in the orbit the planet is located at a given time we need a sixth orbital element, the time of perihelion passage, T.

Table E1. *Some properties of an elliptical orbit*

e	t_{90}	$t_{\frac{1}{2}}$	d_{90}	q
0.1	0.218	0.234	0.990	0.900
0.2	0.187	0.218	0.960	0.800
0.3	0.156	0.202	0.910	0.700
0.4	0.126	0.186	0.840	0.600
0.5	0.098	0.170	0.750	0.500
0.6	0.071	0.155	0.640	0.400
0.8	0.026	0.123	0.360	0.200

Note:

t_{90} and $t_{\frac{1}{2}}$ are given in units of the object's period; d_{90} is in units of the orbit's semi-major axis, a.

In summary, six orbital elements are needed to define an elliptical orbit *and* locate the planet/comet. I prefer to remember them in the following order:

a = semi-major axis Ω = longitude of the ascending node

e = eccentricity ω = angle from ascending node to perihelion measured in the direction of motion

i = inclination T = instant when the planet is at perihelion

As was stated earlier, the trick is to derive these quantities from a set of observations. Much has been written on the subject.

The period, in years, of the planet's orbital motion, P, is now given by $P^2 = a^3$ where a is in AU. Note that the shape of the ellipse has absolutely no effect on the period; skinny or fat matters not. Orbits with the same a-value have the same periods.

Knowing when the planet is at perihelion is fine, but how do we tell where the planet will be a week or a month or a year later? Again, the answer is straightforward but complicated, and will not be given here. It is, however, a simple matter to calculate the planet's velocity, V:

$$V = 29.8 \sqrt{(2/r - 1/a)} \tag{2}$$

where r is the distance from the Sun to the planet, and a is the semi-major axis, both in AU. Again V is in km/s.

Notice that when a planet or a comet is at the half-way point between perihelion and aphelion, r becomes equal to a (see Figure E1), and the object's velocity is the same as if it were in a circular orbit with a equal to the orbit's radius.

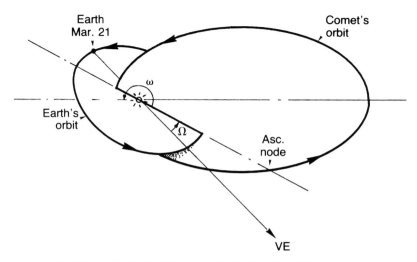

Fig. E2. An elliptical orbit in space. See text for the orbital elements.

But we will want to have some idea as to how far an object moves in a given interval of time. We can estimate this roughly if we know how long it takes the planet to swing through an angle of 90° beginning (or ending) at the perihelion point, and how long it takes to move half-way from perihelion to aphelion, call them t_{90} and $t_{\frac{1}{2}}$. (See Figure E1.) These quantities, which can be calculated from Newton's and Kepler's laws, appear in Table E1 where we give t_{90} and $t_{\frac{1}{2}}$ for various eccentricities in units of the object's period. Also tabulated is the distance, in terms of a, from the Sun to one of the 90° points, d_{90} (also called the *semi-latus rectum* point); and lastly, q, the perihelion distance, again in units of a.

In Figure E2, we have put it all together for a hypothetical comet where we have adopted the following orbital elements:

Comet Hypothet ($2000z_1$)

$a = 4.6416$	$\Omega = 30°$
$e = 0.8$	$\omega = 210°$
$i = 45°$	$T = $ Feb. 30, 2000

We have chosen this particular value of a so that the period will be 10 years (since we know that $P^2 = a^3$). Figure E2 shows the orbit, here drawn in the plane of the paper so as to have the comet's elliptical orbit undistorted by perspective. (To make things fit, the value of a has been reduced to 2.2.) The plane of the Earth's circular orbit is inclined (by 45°) and consequently foreshortened. From the bottom line of Table E1 and since $P = 10$ years, we find that it will take a scant 0.26 years for the comet to swing through an angle of 90° from the perihelion, and it will take it only 1.23 years

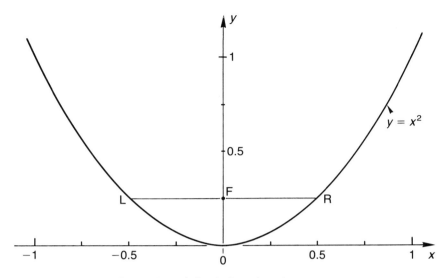

Fig. E3. A parabola. The basic formula is $y = x^2$.

to get half-way out to aphelion. At the 90° points, the comet's distance from the Sun will be $4.64 \times 0.360 = 1.67$ AU. As noted earlier, at the half-way point, the distance from the Sun will be a, and $q = 0.92$ AU.

Parabolic orbits

Whenever a new comet arrives on the scene, there is always a rush to measure accurate positions so that its orbit can be calculated and its future path predicted. Since most new comets come in from the far outer reaches of the solar system, presumably from the Oort Cloud (see Chapter 2), they move in immensely long ellipses, many with eccentricities in excess of 0.99. But the abbreviated part of the ellipse that we see matches almost perfectly the shape of a *parabola*.

The curve shown in Figure E3 is a parabola; it was plotted with the basic formula for a parabola, namely $y = x^2$. The single accessible focus F lies on the vertical axis (as does the other but off in infinity somewhere) and is at a distance of 0.25 from the curve.

In the figure, find the points on the curve at $x = \pm 0.5$, $y = 0.25$; then notice that the focus lies directly between these two points. In other words, the distance from the focus horizontally across to the orbit on either side (the points labelled L and R) is twice the distance from the focus to the perihelion. The line connecting these two side points and passing through the focus is called, as noted for the ellipse, the *latus rectum*. (This Latin term means, incidentally, 'straight side'.)

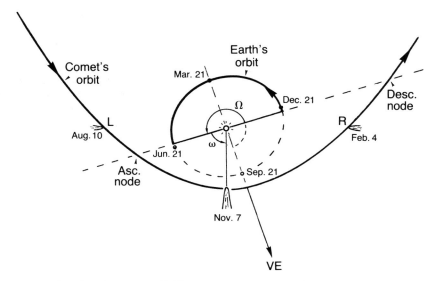

Fig. E4. The orbit of Comet Bradfield (1987s) in space. See text for the orbital elements. For the sake of clarity, q has been drawn to equal 1.08.

Since e is, by definition, unity, only four additional elements are needed to locate a parabolic orbit in solar system space – plus one more to tell where in the orbit the object is at any given moment. The orbital size is now defined by q, the perihelion distance; the remaining elliptical elements – i, Ω, and ω – remain unchanged.

While parabolas come in all sizes, they all have the same identical shape. In this regard they are just like circles. Therefore, since when you have seen one, you've seen them all, I would suggest that you carefully draw a parabola on a clean sheet of centimetre graph paper, and save it. So that the parabola will be a convenient size, make the scale such that a unit distance corresponds to 16 centimetres. (If you only have inch graph paper, then use 6 inches.) The Sun will then be exactly 4 cm from the perihelion. We have done the same thing in Figure E4, but since the scale was determined by the book's printer, the scale has been altered.

Let us now work through a real orbit, showing clearly all the elements. To begin with, trace out a copy of your standard parabola on a sheet of cardboard or fairly stiff paper. We will take as an example the orbit for Comet Bradfield 1987s as published in IAU *Circular* No. 4442. The elements appeared with an ephemeris and this message:

Comet Bradfield (1987s)

The following parabolic elements are from 19 observations Aug. 12–22. The extended ephemeris is mainly for planning purposes.

$$
\begin{array}{ll}
T = 1987 \text{ Nov. } 7.117 \text{ ET} & \omega = 73.672° \\
q = 0.87094 \text{ AU} & \Omega = 267.356° \qquad \Big\} \quad 1950 \\
& i = 34.080°
\end{array}
$$

Since the radius of the Earth's orbit is $1/0.87 = 1.15$ times the perihelion distance of Comet Bradfield, it will have a radius of $4 \times 1.15 = 4.60$ cm on your drawing. If you want to add it, use pencil; you will probably want to erase it later. But the important thing to remember is that your standard drawing of a parabola can be used for any size orbit simply by adjusting the radius of the Earth's orbit to scale. In Figure E4 we have jumped ahead of our story and put in the Earth's orbit in perspective, something that you would normally do as a last step.

The orbit of Comet Bradfield had $\omega = 73.7°$. On your drawing, use a protractor to measure off this angle. Measure it from the perihelion point *counter* to the comet's counter-clockwise motion. (Since i is less than $90°$, we know the comet's motion is direct.) We measure it counter to the comet's direction of motion in order to find out where the ascending node is located. (Remember that ω was defined as the angle from the ascending node measured in the direction of motion, but we have to find out where the ascending node is first.)

All that remains is to find out how the orbit is oriented relative to the celestial sphere. This information is given by the longitude of the ascending node and the inclination, Ω and i.

You can measure off the angle Ω on your diagram as we have done in Figure E4. Now you can find out approximately where the Earth was: first note that *as seen from the Sun*, the Earth is in the direction of the vernal equinox on around September 21. Now mark off the Earth's orbit into four quarters beginning with the September 21 location. These marks will tell you closely where the Earth was on the 21st of December, March, and June. Other dates can be found by interpolation and by remembering that the Earth moves clounter-clockwise almost exactly one degree a day.

Because the inclination of Comet Bradfield's orbit is not too large, the orbits of both the Earth and the comet can be drawn on the same diagram without much distortion. Thus, the pencilled circle that you made represents the Earth's orbit reasonably well. But notice that, in truth, the comet's orbit 'hinges' up on the line passing through the two nodes and the Sun.

For those of us with little or no artistic talent, there is, happily, another, perhaps even better way of showing the two orbits. Along the line of nodes, cut a 9.20-cm slot with its centre at the Sun. On another piece of stiff paper, cut out a circle of this diameter and insert it (the Earth's orbit) in the slot; its centre should be at the Sun. Now tilt the two to make a $34.1°$ angle between the two pieces of paper. If you have the comet coming up through the Earth's orbital plane at the ascending node, you have the finished representation. (A little triangle of paper having a $34°$ angle inserted and glued to both planes can keep the alignment permanently fixed.)

As for the comet's motion, its orbital speed (V in km/s) is given by equation (2), but since for a parabola $a = $ infinity, this formula simplifies to

$$V = \frac{42.1}{\sqrt{r}}. \tag{3}$$

Table E2. *The time taken for a comet to move 90° from perihelion*

q	t_{90}	q	t_{90}
0.2	10 days	1.0	110 days
0.4	28	1.5	201
0.6	51	2.0	310
0.8	78	3.0	570

Also, the comet will be at the two *latus rectum* points 90° from perihelion t_{90} days, where $t_{90} = 109.6q^{\frac{3}{2}}$ (q in AU), before and after T; Table E2 gives values for comets with different perihelion distances. For Comet Bradfield, this quantity is 89 days; therefore, it would be at the *latus rectum* points on August 10, 1987 and February 4, 1988.

Your diagram shows clearly how Comet Bradfield moved with respect to the Earth: it came up from beneath (south of) the Earth's orbit and cut through the ecliptic sometime in September (actually on the 9th). The Earth was up ahead of the comet at the time. For almost the entire next six months, the comet remained north of the ecliptic passing directly over the Earth's orbit as it neared perihelion. By the time Comet Bradfield reached perihelion on November 7, the Earth had swung around in its orbit so that it was moving away from the comet. By year's end, the comet was moving rapidly away from both the Sun and the Earth.

One can easily put a scale of right ascensions on the diagram since we know that the Sun is at RA $= 12^h$ on September 21, 18^h on December 21, and so on around the orbit. Declinations can also be estimated if we remember that the ecliptic plane is mapped on the sky as a wavy line 23.5° north of the celestial equator at RA $= 6^h$, 23.5° below the equator at RA $= 18^h$, and crossing the equator at RA $= 0^h$ and 12^h.

According to the ephemeris, on November 1 Comet Bradfield crossed the celestial equator at RA $= 17^h 28^m$. Its distance from the Sun, r, equalled 0.878 AU and its distance from the Earth, Δ, equalled 1.130 AU. How closely could you predict this from your three-dimensional model?

It may take a little time to puzzle out the first few orbit models, but after some practice, you will be able to sketch out quickly the orbit of each newly discovered comet and to tell at a glance how the comet moves relative to the Earth and when the viewing will be best. Better yet, you will be able to visualise everything in your mind.

Appendix F
Discoveries during the 1980s

Initial remarks

This appendix, devoted primarily to the discoveries made during the 1980s, contains numerous graphical and tabular presentations of the discovery circumstances. Anyone seriously interested in joining in the astro-search should devote some time to a careful study of this information since it shows the who–what–when–where–how connected with each discovery. Like in many games, to emulate is to succeed.

Emphasis will be on what I call 'amateur discoveries'. The distinction between amateur or professional, sometimes rather hazy, has been made by me on the basis of what information I have and at times may seem rather arbitrary or just plain wrong. Some of the discoveries that you will see listed were made by pros, but they *could* have, or even *should* have been made by amateurs. Furthermore, despite a conspicuously professional background, I persist in calling myself an amateur (born-again). My reasons are two: I use equipment well within the price range of your average *aficionado*, and my motivation comes entirely from the heart, not the pocketbook.

The order of subject matter, namely the objects discovered, will follow that of Chapters 2, 3, and 4. The information comes mainly from the original discovery announcements given in the IAU *Circulars*, but other sources will be noted in each section.

Comets

First off, I should note that Don Machholz has carried out an excellent detailed study of the discovery circumstances of comets in his newsletter *Comet Comments*. Some of the graphs in this section were suggested by his splendid efforts, and some of the information was skimmed from his articles. Michael Rudenko's detailed 'Catalogue of Cometary Discovery Positions' likewise provided a number of vital facts and figures. It appeared in *The International Comet Quarterly*, a semi-official source of information, which describes itself as 'a journal devoted to news and observations of comets'. In the UK, the monthly publication *The Astronomer* likewise provides the latest words on comet discoveries and observations, mainly British. Information on these (and other) journals and how to acquire them will be found in Appendix H.

We begin with Table F1 which lists the amateur comets, 37 of them, discovered by 44 individuals and (jointly) one piece of hardware, IRAS (the Infrared Astronomy Satellite), in the 1980s. A few comments are noted below.

When more than one discoverer is listed (first column) and the discoveries were made independently, separate listings of the telescope information (Tel.) are given for each named individual. Thus, Churyumov and Solodovnikov worked together (Comet 1986 IX), but Skorichenko and George (Comet 1989 e_1) worked separately. The Local Time (LT) is calculated from the observer's longitude that sometimes had to be 'guesstimated'. The comet's magnitude, m, is taken directly from the IAU *Circulars*, and it is usually on something approximating the visual system. In the 'Tel.' column, following the diameter of the objective in centimetres, p, v, and B indicate photographic, visual, and binocular observations, respectively.

The comet's coordinates are given only once per comet, except for Comet IRAS–Araki–Alcock (1983 VII) which initially was thought, by IRAS scientists, to be an asteroid. It wasn't until the following week that the true cometary nature was discovered. Similarly, the comet's solar distance, r, the angular distance between comet and Sun, *C-S*, the rate of motion in minutes of arc per hour, and the direction of motion are only given once per comet since they usually change little at first appearance. The last two columns give the age of the Moon in days at time of discovery, and the number of the IAU *Circular* on which the discovery was first announced.

Using the data of Table F1, one could invent dozens of informative graphs that would give a better 'feeling' for the discovery circumstances. Here we present some of the more obvious correlations. For the would-be comet discoverer, probably the most useful single piece of information is where in the sky the comet was found or, to be more specific, what were the comet's altitude and azimuth at time of discovery? The first figure presents these data, and immediately it is seen (Figure F1) that the most favoured areas in the sky are located due east and due west, each spreading over $\pm 45°$ of azimuth, and at altitudes between 5° and 45°. All but five comets were discovered in these 'boxes'. A perusal of Table F1 confirms what should be obvious: the comets in the east are seen before sunrise; those in the west after sunset. (The one comet at 79° altitude, Comet Furuyama 1988 IV, was discovered photographically a little over an hour after local midnight – presumably by accident.)

Figure F1 would be more instructive if it could somehow indicate for each comet how far below the horizon the Sun was, but that would complicate matters greatly. But we can illustrate in general terms how long before sunrise or after sunset the comet discoveries were made. Figure F2 shows the altitude of the Sun (always negative, of course) in the form of a histogram. Evening comets are cross-hatched; morning dotted. The conclusion to come away with is that more than two-thirds of the discoveries were made when *the Sun was less than 30° below the horizon*, and almost half less than 20°. At temperate latitudes, a $-30°$ altitude for the Sun corresponds to between $2\frac{1}{2}$ and 3 hours after sunset or before sunrise.

Table F1.37 'Amateur' comets discovered between January 1, 1980, and December 31, 1989

Name(s)	Desig.	Discov. UT (yymmdd.hh)	LT (h)	m	Tel. (cm)	RA (2000) (h)	Dec. (2000) (°)	r (AU)	C-S (°)	Motion (arcmins/h)	Dir.	Moon age (d)	IAUC
Černis–Petrauskas	80 IV	800731.17	21.5	9	11B 8B	11.88	+32.4	1.1	45	2.9	E	19	3498
Meier	80 XII	801106.03	21.5	10	40v	18.13	+42.1	1.6	75	1.5	SSW	28	3535
Bradfield	80 XV	801217.18	03.2	6	3B	16.38	−36.4	0.3	21	5.1	ENE	10	3554
Panther	81 II	801225.18	19.7	10	20v	18.82	+39.0	1.7	63	0.8	NNE	19	3556
Austin	82 VI	820618.16	03.5	10	15v	4.12	−40.0	1.1	68	0.9	NE	26	3705
Sugano–Saigusa–Fujikawa	83 V	830508.18	03.3	7	15v 20v 12B	1.60	+39.7	0.5	29	1.6	WNW	25	3803 3803 3803
IRAS–Araki–Alcock	83 VII	830425.21	—	—	—	19.17	+48.7	1.1	93	3.5	NNW	21	3796
		830503.15	23.8	7	7v	18.95	+52.6					21	3796
		830503.22	22.0		8B								3796
Černis	83 XII	830719.00	05.0	12	48v	2.77	+12.0	3.3	73	1.2	S	9	3840
P/Bradfield[1]	83 XIX	840107.17	03.8	11	25v	15.93	−46.9	1.4	47	2.6	SE	5	3907
P/Takamizawa[2]	84 VII	840730.12	21.5	10	12B	21.25	−18.5	1.9	171	0.6	SSW	3	3964
Austin	84 XIII	840708.17	05.0	8	15v	4.88	−38.7	0.6	69	3.4	E	11	3957
Meier	84 XX	840918.01	19.7	12	40v	15.18	+11.1	1.0	53	2.1	SW	23	3991
Levy–Rudenko	84 XXIII	841114.03	19.3	9	40v	18.82	+9.9	1.2	59	1.5	N	21	4007
		841115.01	20.2	10	15v							22	4007
Machholz	85 VIII	850527.11	03.2	9	25v	0.88	+15.4	0.9	49	4.1	ENE	7	4067
P/Ciffreo[3]	85 XVI	851108.02	02.7	10	90p	4.59	+23.5	1.7	155	0.7	NNW	24	4135
P/Machholz[4]	86 VIII	860512.11	02.5	11	13B	0.73	+38.9	0.7	39	3.7	WNW	3	4214
Churyumov–Solodovnikov	86 IX	860714.19	00.3	13		21.93	−12.1	2.7	115	1.8	SW	7	4233

Levy	86 XVII	870105.12	04.7	10	40v	17.35	+10.4	1.0	97	0.9	SW	6	4295
Terasako	86 XVIII	870124.09	18.5	8	15B	23.32	−30.2	0.9	40	4.5	ENE	24	4303
Sorrells	87 II	861101.08	23.5	12	40p	5.70	+26.9	2.3	133	3.5	W	28	4267
Nishikawa–	87 III	870119.11	20.2	9	15v	0.08	+5.3	1.3	65	3.5	SW	19	4300
Takamizawa–		870120.09	18.5	9	15B							20	4300
Tago		870120.10	19.2	9	15B							20	4300
Levy	87 XXI	871011.03	19.7	9	20v	14.59	+17.3	0.9	33	3.7	E	19	4468
Rudenko	87 XXIII	870821.03	21.7	10	15v	14.17	+33.8	1.1	61	1.8	WSW	27	4440
Bradfield	87 XXIX	870811.10	19.7	10	15v	14.24	−23.6	1.8	81	0.9	NE	17	4431
Levy	87 XXX	880319.12	04.7	11	40v	21.59	+16.9	1.1	39	1.4	NE	3	4566
McNaught	87 XXXII	871018.10	20.0	9	4p	14.45	−52.2	1.3	44	2.4	NE	25	4473
Ichimura	88 I	871122.12	21.7	9	12B	4.01	−19.1	1.3	144	7.6	SSW	1	4494
Furuyama	88 IV	871123.16	01.3	12	30p	5.35	+25.5	2.2	160	3.1	SW	3	4499
Liller	88 V	880111.01	20.5	10	20p	23.90	−28.0	1.9	59	1.7	N	22	4527
Machholz	88 XV	880806.11	03.0	9	12B	4.73	+0.8	0.9	66	5.6	E	25	4636
Yanaka	88 XX	890101.18	03.0	11	15B	13.80	+9.5	2.0	81	1.5	ENE	24	4697
P/Bradfield[5]	88 XXIII	890106.12	21.5	12	25v	21.22	−56.2	0.9	41	4.2	ENE	28	4703
Yanaka	88 XXIV	881229.20	05.2	9	15B	16.63	+1.0	0.8	37	4.3	SW	22	4696
Okazaki–	89 r	890824.12	21.2	11	25p	15.53	+34.2	1.7	75	1.0	WSW	23	4841
Levy–		890825.05	21.5	11	40v							24	4840
Rudenko		890826.04	23.0	11	25v							25	4840
Aarseth–	89 a$_1$	891116.17	17.8	8		16.32	+28.6	1.0	49	1.9	S	19	4907
Brewington		891117.00	18.2	8	40v							20	4907
Austin	89 c$_1$	891206.14	01.7	11	20v	0.91	−62.0	2.4	83	0.9	NNW	8	4919
Skorichenko–	89 e$_1$	891217.17	19.5	11	15v	19.89	+25.6	2.2	59	1.1	ENE	20	4929
George		891218.00	19.5	11	40v							21	4925

Notes:

(1) $P=151^y$, (2) $P=7.23^y$, (3) $P=7.23^y$, (4) $P=5.24^y$, (5) $P=81.0^y$

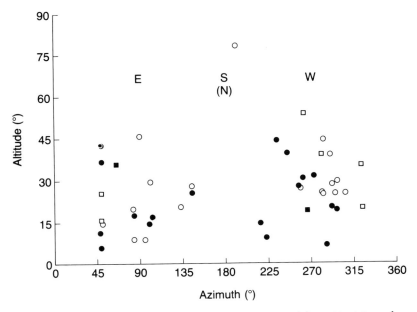

Fig. F1. The altitudes and azimuths of the amateur comets of the 1980s at time of discovery. Squares represent southern hemisphere discoveries; open figures signify fainter than magnitude 10.0.

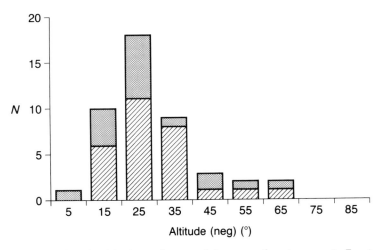

Fig. F2. The altitude of the Sun at the time of discovery of amateur comets. Evening discoveries are cross-hatched, morning dotted.

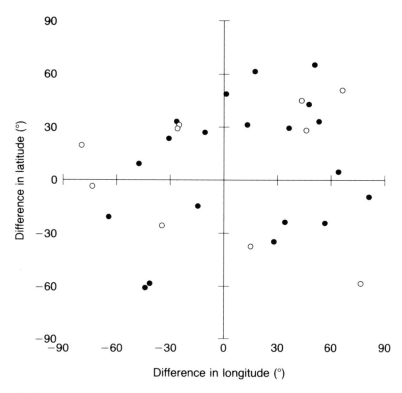

Fig. F3. The locations of just-discovered amateur comets relative to the Sun. Comets fainter than mag. 10.0 at discovery are indicated by open symbols. Six with $b > 120°$ are omitted.

In this connection, we should note that by definition, 'astronomical twilight' begins or ends when the Sun has reached an altitude of $-18°$. At this point the sky is dark enough to suit the most pernickety professionals. When the Sun is $12°$ below the horizon, we have 'navigational twilight', and many visual comet seekers start – or finish – then.

One might expect fainter comets to be discovered higher in the sky on the average, partly because comets are generally brightest when near the Sun, and partly because the increased atmospheric extinction at low altitudes makes it difficult to pick up fainter objects. However, except perhaps at the very lowest altitudes, Figure F1 does not show this. (The open circles and rectangles represent comets fainter than mag. 10.)

There are several other ways to show the relationships of the discovery positions with respect to the Sun. Figure F3 shows the positions of all but six of the amateur comets plotted in the ecliptic latitude–longitude system relative to the Sun which is

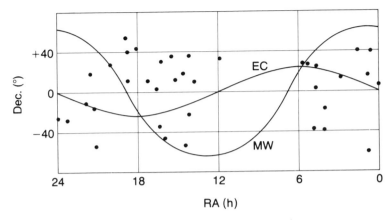

Fig. F4. A map of the sky showing where the newly discovered amateur comets were located. The locations of the ecliptic (EC) and the Milky Way (MW) are indicated.

located at the centre. Notice that except for the half-dozen odd-balls found far from the Sun (all but one were close to the ecliptic), at time of discovery the comets showed no strong positional tendencies – and most would fit nicely within a circle 60° in radius centred on the Sun.

Figure F4 is nothing more than an equatorial map of the sky showing a few coordinates and the locations of the ecliptic and the Milky Way. Again we see that amateur comets are no more often discovered near the ecliptic than near any other arbitrary line in the sky. Also, it seems that the Milky Way has little effect on discoveries: comets seem to be picked up there as frequently as any other place in the sky.

When I first plotted this map, I was startled that there had been no comets discovered in the 6-hour zone centred on about 8.8 hours RA. Statistically, the chances of this happening by chance are about one in 2000. To be sure, there is a bothersome clustering of galaxies around an RA of 12 hours and north of the equator, but at lesser RAs, there are good hunting grounds. A similar chart that Don Machholz published in his *Comet Comments*, a map that includes comets going back to 1975, does not show this vacancy. Therefore, I have to conclude that during the 1980s this dearth of comets just happened by chance, a statistical 'glitch' as it were.

The reason that comets are found all over the sky, and not just near the ecliptic like asteroids, is, of course, that most of the new discoveries are long-period comets travelling in near-parabolic orbits with high inclinations well represented, as Figure F5 shows. These data were taken from Marsden's *Catalogue of Cometary Orbits* (1989 edition) Notice, on the other hand, that short-period comets (dotted columns) tend to stay close to the ecliptic. These behavioural characteristics (among other things) have led to the suggestion that all the short-period comets were originally long-

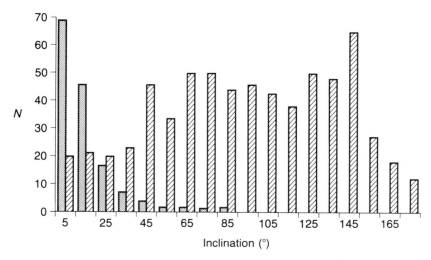

Fig. F5. The inclinations of the orbits of all comets with well-determined orbits. Long-period comets are indicated by cross-hatch; short-period comets by dotted areas.

period comets with low orbital inclinations but at some time passed close to one of the major planets and had their orbits altered.

Figure F6 shows the orbital periods of all well-observed short-period comets listed in the Marsden catalogue; the fact that there is a strong peak at 6 years is not accidental; Jupiter's powerful gravity makes it that way.

We now turn to the visibility factors involved, beginning with the discovery magnitudes listed in Table F1. Figure F7 shows a histogram of these data, not just for the amateur comets (cross-hatched columns) but also for the Palomar comets (dotted). The amateur comets cut off rather sharply at magnitude 12 (the one comet at mag. 13 could easily be classified as a professional comet), and with two exceptions, the Palomar comets begin abruptly at 13. However, don't forget the words of Chapter 2 and Figure 2.6: only one Palomar comet, the one at mag. 11, was discovered closer than 100° from the Sun. In other words, as we concluded, the best way to discover comets might just be to search to around 11th to 13th magnitude at elongations of about 90° from the Sun.

As most of the successful visual comet seekers emphasise, dark skies are important, and one should get out of town to search the skies. But the Moon can also be a problem, as is well shown in Figure F8. The full-Moon hiatus is conspicuous. Equally importantly, note the string of morning discoveries made around the time of first quarter Moon. Remember that the first quarter Moon sets around midnight, leaving the morning skies free of moonlight. Similarly, beginning a few nights after full Moon, there is time between end of evening twilight and moonrise to do some dark sky comet hunting. And from the rash of discoveries that comes when the

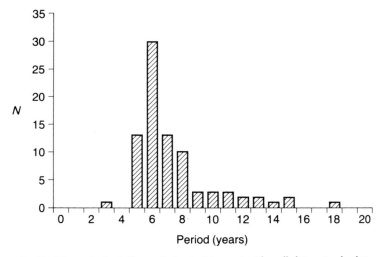

Fig. F6. The periods of all comets (up to 20 years) with well-determined orbits.

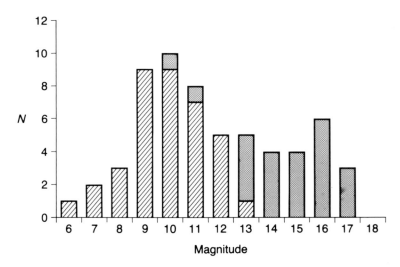

Fig. F7. The magnitudes of comets at time of discovery. The amateur comets are cross-hatched; the 'Palomar' comets dotted areas.

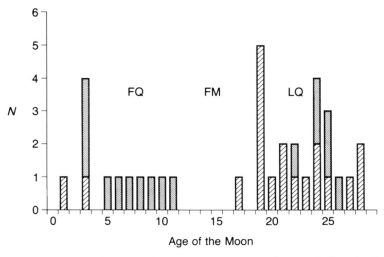

Fig. F8. The age of the Moon when amateur comets were discovered. Cross-hatched columns indicate evening discoveries; dotted areas morning discoveries. The principal phases of the Moon, denoted by FQ, FM, and LQ, are indicated.

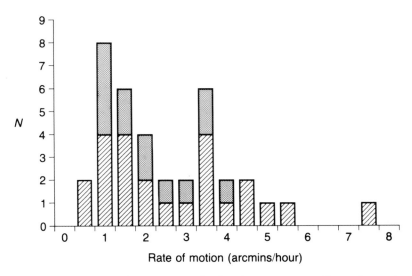

Fig. F9. The angular rate of motion across the sky of amateur comets. Comets fainter than mag. 10 are indicated by dotted columns.

Moon's age is 19 days, it would seem that is the time when many comet seekers begin seriously scanning the sky. But Tetsuo Yanaka didn't let the Moon stop him. (See Chapter 5.)

The last figure in this section is concerned with how fast, in angular units, the newly discovered amateur comets were moving across the sky. In other words, how long must one wait after finding a suspicious fuzzy patch before motion should become evident? Figure F9 tells us that while there are a few fast movers, you should assume that in an hour your candidate will move no more than one or two minutes of arc. The comet's brightness or angular distance from the Sun gives few clues to its angular motion. A faint comet could be a monster many AU from the Sun and Earth, or a measly one up close. The first would creep along slowly; the latter very likely would zip by briskly. Even a monster comet up close could be heading right at the Earth in which case (1) it would show no angular motion, and (2) please call me if you find one – immediately.

Novae

Almost all the data in this section were taken from the IAU *Circulars*; a few additional data for novae have been taken from Duerbeck's excellent *A Reference Catalogue and Atlas of Galactic Novae*.

Again we start with a tabulation of discoveries during the 1980s; in Table F2 are listed the discoveries of all galactic novae. You can see that five became naked-eye objects (mag. 6.5 or brighter), and the faintest was found by a Soviet astronomer with a peak magnitude of 15.0. Note that a nova might have been found many days after it reached its actual brightest magnitude, in which case the listed peak magnitude would refer just to the maximum observed brightness. In a few cases, photographs taken serendipitously before discovery were examined after the fact, and the nova was found to be brighter than when discovered, and that is the magnitude listed. Of course sometimes (never in the 1980s to my knowledge) the nova gets discovered *before* the maximum was reached.

Novae are regularly found in other galaxies, mainly the Magellanic Clouds where they peak around 11th magnitude. Southern hemisphere nova hunters have it lucky. The nearest northern hemisphere galaxy of any respectable size is M31 in Andromeda, and there novae reach mag. 16 under good conditions (i.e., no interstellar extinction).

In the first graph (Figure F10), we show where in the sky the 35 novae were found. The star that some call a dwarf nova, VY Aqr, was seen to explode four times; it gets plotted only once. The concentration to the Milky Way is striking, especially around 18 hours RA; Figure F11 shows this region expanded. The reason for this congregation was explained in Chapter 3, and it is, of course, that the dense centre of our Galaxy is in this direction. The location of the galactic nucleus is labelled 'GC' and marked with an 'X'.

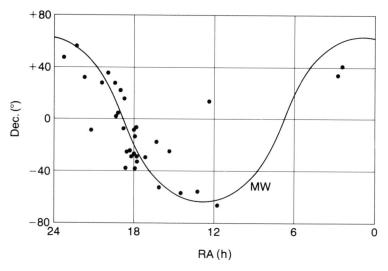

Fig. F10. The location of the novae of the 1980s in the sky. The Milky Way (MW) is indicated by the curved line.

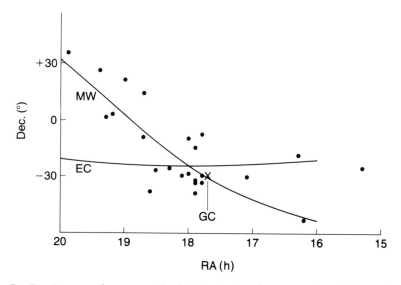

Fig. F11. Location of novae near the direction to the galactic centre, labelled GC and marked with an 'X'. The locations of the Milky Way (MW) and the ecliptic (EC) are also shown.

Table F2. *Galactic novae: January 1, 1980, to December 31, 1989*

Designation	Discvr.	Date of discov.	Type of observ.	RA (1950) (h m)	Dec. (° ')	Peak mag.	l	b	IAUC
Vir 1980	Gonzalez	11 Jul.	p	12 23.1	+13 10	14	279	+75	3500
Sgr 1980	Honda	28 Oct.	p	18 16.5	−24 45	9.0	7	−05	3533
Cyg 1980	Honda	29 Nov.	p	21 40.8	+31 14	10	82	−16	3546
CrA 1981	Honda	2 Apr.	p	18 38.6	−37 34	7.0	358	−14	3590
Aql 1982	Honda	27 Jan.	p	19 20.8	+02 24	6.5	39	−06	3661
Sgr 1982	Honda	4 Oct.	p	18 31.5	−26 28	8.0	7	−08	3733
Mus 1983	Liller	18 Jan.	p	11 49.6	−66 56	7.2	297	−05	3764
Sgr 1983	Wakuda +Ogura	19 Feb.	p	18 04.7	−28 50	9.5	3	−04	4119
Ser 1983	Wakuda	21 Feb.	p	17 53.2	−14 01	8	14	+06	3777
Cep 1983*	Honda	1 Jun.	p	22 12.0	+56 46	7.5	103	00	3821
Lib 1983	Gonzalez	10 Aug.	p	15 17.0	−24 50	9.0	341	+27	3854
Tri 1983	Kurochkin	11 Sep.	p	2 42.2	+33 19	15.0	149	−24	3869
Nor 1983	Liller	19 Sep.	p	16 09.9	−53 12	9.4	330	−02	3869
VY Aqr*	Fujino +McNaught +Bortle	28 Nov.	—	21 09.5	−09 02	10.3	42	−35	3896
Vul 1984	Wakuda	27 Jul.	p	19 24.1	+27 16	6.3	61	+05	3963
Sgr 1984	Liller	25 Sep.	p	17 50.5	−29 02	9.7	1	−02	3995
Aql 1984	Honda	2 Dec.	p	19 14.1	+03 38	10	39	−04	4020
Vul 1984	Collins	22 Dec.	B	20 24.7	+27 21	5.1	68	−06	4023
RS Oph*	Morrison	26 Jan.	v	17 47.5	−06 42	5.2	20	+10	4030
Sco 1985	Liller	24 Sep.	p	17 53.3	−31 49	10.5	359	−04	4118
Cen 1986*	Liller	13 Jan.	p	13 17.7	−55 35	7.5	307	+07	4180

Table F3. *Galaxies with supernovae mag. 15 or brighter in the 1980s*

NGC gal.	M No.	SN m_{peak}	Gal. type	SN type	Date of discov.	First discoverer
521#		15.0	SBc	—	19 Aug. 82	Lovas
991#		13.8	Sc	I	28 Aug. 84	Evans
1187*		14.4	SBbc	I	24 Oct. 82	Muller
1316#*		12.5	Sap	I	30 Nov. 80	Wischnejewsky
1316#*		12.7	Sap	—	10 Mar. 81	Evans
1332*		14	S0	—	29 Mar. 82	Wischnejewsky
1365#*		13.5	SBb	II	25 Nov. 83	Evans
1433#*		13.5	SBb	II	10 Oct. 85	Evans
1448#*		14.5	Sc	II	6 Oct. 83	Evans
1532#*		13.5	Sab	I	24 Feb. 81	Evans
1559#*		13.2	SBc	II	27 Jul. 84	Evans
1559#*		13.5	SBc	II	7 Oct. 86	Evans
1667#		15	Sc	I	11 Dec. 86	Pennypacker
2227		14	SBcd	I	24 Nov. 86	Pennypacker
2268		14	Sbc	—	12 Feb. 82	Wild
2336*		14.2	SBbc	I	16 Aug. 87	Patchick
2715		15.0	Sc	I	21 Sep. 87	Lovas
2748		14.5	Sc	—	25 Jan. 85	Schildknecht
3044#		14	Scd	—	13 Mar. 83	Metlov
3169#		14.0	Sb	II	26 Mar. 84	Evans
3227#		12.0	Sb	I	4 Nov. 83	Pronik
3336		15.0	S	—	20 Dec. 84	Wischnejewsky
3367#*		14.0	SBc	I	4 Feb. 86	Evans
3667#*	66	12.2	Sb	I	30 Jan. 89	Evans
3646		14.5	Sbc	—	30 Jun. 89	Mikolajczak
3675*		13	Sb	—	2 Dec. 84	Ikeya
4045#		13	Sbc	II	17 Jan. 85	Horiguchi
4051*		13.5	Sbc	—	11 May 83	Tsvetkov
4220		14.5	Sa	I	30 Jun. 83	Wild
4254#*	99	14.0	Sc	II	17 May 86	Pennypacker
4302#		14.5	Sc	II	13 Apr. 86	Candeo
4374#*	84	14	E1	I	31 Jun. 80	Rosker
4419#		12.8	SBab	I	4 Jan. 84	Kimeridze
4451#		14.5	S0	II	21 Mar. 85	Horiguchi
4536#*		12	E0/S0	I	2 Mar. 81	Tsvetkov
4579#*	58	13.5	Sab	II	18 Jan. 88	Ikeya +
4579#*	58	12.2	Sab	I	28 Jun. 89	Kimeridze

Table F3 (*cont.*)

NGC gal	M No.	SN m_{peak}	Gal. type	SN type	Date of discov.	First discoverer
4651#		15	Sc	II	28 Jul. 87	Pennypacker
4699#*		14	Sab	—	6 Jun. 83	Wischnejewsky
4716#		15	S0	I	30 May 81	Wischnejewsky
4753#*		13	S0p	I	5 Apr. 83	Okasaki +
4874		15	E	—	2 Jun. 81	Lovas
5033*		12.5	Sbc	II	13 Jun. 85	Metlova
5128#*		11.7	S0 + Sp	I	3 May 86	Evans
5236#*	83	11.2	SBc	—	3 Jul. 83	Evans
5485		15	S0p	I	14 Dec. 82	Lovas
5746#		13	Sb	I	11 Jul. 83	Pellegrini
5850#*		15	SBb	II	24 Feb. 87	Evans
5854#		15.0	Sa	I	20 Mar. 80	Faber
6907		15.0	SBbc	—	29 May 84	Gonzalez
6946*		11	Sc	II	28 Oct. 80	Wild
7184		14	Sb	I	20 Jul. 84	Evans
7448#		13.5	Sbc	—	8 Oct. 80	Inasaridze
7606#*		13.8	Sb	II	14 Dec. 87	Evans
Other galaxies						
IC 121#		14.0	—	—	29 Aug. 84	Lovas
IC 1731		14	Sc	I	3 Oct. 83	Wild
LMC#*		2.8	SB	II	24 Feb. 87	Shelton +
ANON		15	—	—	22 Nov. 82	Wild
ANON		14.5	Sc	—	3 Feb. 89	Wild
MCG		15	Sa	—	10 Aug. 81	Wischnejewsky
MCG		14	—	I	20 Apr. 87	Metlova
Mk 516		15	—	I	18 Oct. 85	Cohen
ZW		15	Sa	I	17 May 88	Maury
UGC		15	—	II	4 Oct. 89	Mueller

Notes:

A # symbol means that a photograph of the galaxy appears in one of the first two volumes of the Lopez Alvarez *SN: Photographic Atlas of Galaxies for Supernova Search.* An asterisk signifies that a chart of the galaxy appears in the Thompson–Bryan *Supernova Search Charts.*

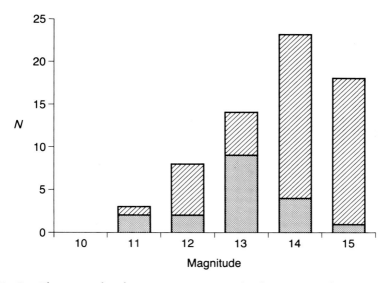

Fig. F14. The magnitudes of supernovae at maximum brightness. Evans' discoveries are indicated by dotted columns.

In this section we present three figures. The first, Figure F14, shows the peak magnitudes of the SNe with Evans' 18 galaxies indicated with dotted columns. The drop at mag. 15 comes because only SNe that reached an estimated mag. 15.0 are listed; a SN that got only to 15.5, for example, is omitted. And, of course, one SN lies off the scale: SN 1987A in the LMC.

Note that Evans discovered or co-discovered *exactly half of all* SNe that reached brighter than mag. 14, a remarkable record, especially considering that some of the galaxies, like NGC 6946, are perilously far north for this Australian.

The histogram in Figure F15 shows in what kind of galaxy the SNe were found. Evans' discoveries are again indicated by dotted columns. Here, however, there are several strong biases in effect − two, at least. First, it is much easier to spot a new stellar object in a washed-out spiral, like an Sc, than a compact elliptical galaxy or one with tightly wound spiral arms (S0 and Sa). It is a little surprising that Evans found so many of his SNe in Sb galaxies which have conspicuous arms. (M31 in Andromeda is an Sb.) However, in the *Shapley−Ames Catalogue*, there are more Sbs and Scs, when added together, than anything else. Figure F16 shows this; it shows the number of Shapley−Ames galaxies of each type.

But notice that in general, the frequency distribution of galaxy types that produced discoverable SNe (Figure F15) closely resembles that given in the Shapley−Ames (Figure F16). In other words, SNe are seen just about as frequently in the S0s as they are in the Scs, only there are are more Scs than S0s. As for the Es, it seems clear that SNe are truly harder to find in these compact, featureless galaxies.

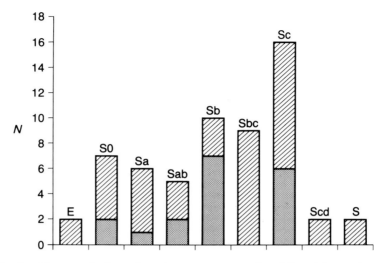

Fig. F15. The type of galaxy where the supernovae were found. Evans' discoveries are indicated by dotted columns.

The conclusions are that if you just want to find a SN, then watch Sbs and Scs that are big in absolute terms (i.e., have lots of stars), and are relatively nearby (i.e., their SNe will be relatively bright). Remember, too, from Chapter 3 that Type I SNe can appear in any type of galaxy, but Type IIs usually are seen just in Sb and Sc galaxies. Because Type I SNe average about a magnitude brighter than Type II in absolute terms, this visibility advantage tends to compensate for the higher surface brightness of the S0 and Sa galaxies. Also, Type Is usually do not appear in spiral arms, but rather between them or out beyond them.

Finally, I should mention the prolific galaxies, the ones that at least seem to produce more SNe than the average for their size and type. In Table F4 are listed those galaxies in which three or more SNe have been discovered as of 1990, together with their coordinates – just in case you want to rush out tonight and start observing. The visual magnitudes of the galaxy and their coordinates are taken from the very useful *NGC 2000.0*, edited by Roger Sinnott. The expected magnitudes of Type I SNe (Expt. m_{SNI}) come either from Thompson and Bryan or from my own estimate. But in both cases I have added a magnitude to take into account interstellar extinction.

Note the clear northern hemisphere bias in Table F4. The reason for this is that, until recently, nearly all SN searches were made with northern telescopes. This fact make M83 especially interesting: who knows how many SNe really have occurred over the years in this bright southern barred spiral?

A regular programme of observation of the galaxies listed in Table F4 should turn up a SN every couple of years, at least. As Table F3 shows, three produced *discovered*

Table F4. *Ten prolific galaxies*

NGC gal.	M no.	No. SNe	Gal. type	Gal. m_v	Expt. m_{SNI}	RA (2000) (h m)	Dec. (° ')
2276		3	Sc	11.4	16	7 27.0	+85 45
2841*		3	Sb	9.3	14	9 22.0	+50 58
3184*		3	Sc	9.8	11	10 18.3	+41 25
4157		3?	Sbc	12.0	15	12 11.1	+50 29
4254*	99	3	Sc	9.8	13	12 18.8	+14 25
4303*	61	3	Sc	9.7	13	12 21.9	+4 28
4321*	100	4	Sc	9.4	12	12 22.9	+15 49
5236*	83	6	SBc	7.6	11	13 37.0	−29 52
5457*	101	3	Sc	9.6	12	14 01.5	+54 36
6946*		5	Sc	8.9	11	20 34.8	+60 09

Note:

* Galaxies appearing in Thompson and Bryan's *Supernova Search Charts.*

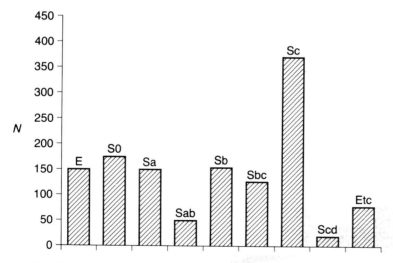

Fig. F16. The number frequency of galaxy types in the Shapley-Ames catalogue.

Table F5. *Asteroids by the thousands*

Asteroid number	Name	Year of designation
1	Ceres	1801
500	Selinur	1903
1000	Piazzia	1923
1500	Jyvaskyla	1938
2000	Herschel	1976
2500	Alascattalo	1981
3000	Leonardo	1984
3500	Kobayashi	1986
4000	Hipparchus	1988
4500	Pascal	1990
5000	IAU	1991

SNe during the 1980s. How many others occurred while the Sun was in the way or while nobody was looking?

This brings us to the question of how often, on the average, a supernova can be expected to appear in a galaxy. Estimates of this quantity, called SNU for Supernova Unit, vary widely, but average something like once every 50 years for a galaxy the size of M31. If the world had about three more Bob Evanses (two in the north, one down south to help out the good Reverend), a better SNU value could be obtained.

Asteroids

Asteroid data have been taken from various sources besides the IAU *Circulars*: the *The Minor Planet Bulletin* of ALPO, the Association of Lunar and Planetary Observers; the *Minor Planet Circulars* (MPC) of the Minor Planet Center at the Smithsonian Astrophysical Observatory; and *Ephemerides of Minor Planets for 1991*, published in Leningrad by the Institute of Theoretical Astronomy of the Soviet Academy of Science. I should also mention the highly enjoyable but no longer published *Tonight's Asteroids* (most ably replaced by Brian Warner's *Minor Planet Observer*). While the MPC is the official journal of asteroid observations, designations, and orbit information (published on behalf of Commission 20 of the IAU), I don't recommend that you run right out and start subscribing. Each year it comes to several thousand pages full of coordinates and observation times, and much of it makes for dull reading – unless you have a fascination with position measurements and orbit determinations. It is also available in other formats for the well-computerised amateur.

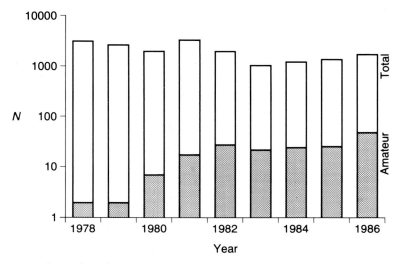

Fig. F17. The number of asteroids discovered per year (adopted from Marsden's article in 'Stargazers'). Note that the vertical scale is logarithmic.

Main belt asteroids

Each year there are dozens of 'normal' asteroids seen for the first time, and dozens more that were seen at other oppositions but not observed diligently enough to enable a decent orbit to be derived. If you have read the accounts of Messrs Manning, McNaught, and Seki in Chapter 5, you know that amateurs are very much involved in the search, not so much for the strange minor planets (the Earth-threateners) — those usually require large apertures and excellent skies to spot — but rather in picking up main belt asteroids as they near opposition.

Table F5 shows at a glance the dizzying increase in the discovery rate of minor planets. Currently, Marsden and Co. have been adding, on the average, more than one asteroid number a day. Where and when is it all going to stop? It is difficult to say. So far the available computer power is keeping up with the enormous task of keeping track of observations and grinding out orbits and ephemerides; as is the manpower needed to feed new observations into the databank.

But why worry about kilometre-sized rocks orbiting over an AU from the Earth? Partly, I suppose, because they're there, but also because collisions can occur, both between asteroids and with space probes that pass through the belt. More interestingly, space probes are now rendezvousing with some of these miniature worlds.

And so just where do the amateurs fit in? Right in the thick of it. Figure F17 shows the remarkable record from 1978 until 1986. While the total number of annual discoveries actually declined slightly, the amateur finds went from an almost negligible two (in 1978) to 48. What is even more remarkable is that there had been *zero* amateur discoveries between 1924 and 1978. The Japanese amateur Takeshi Urata started it all, and he was quickly joined by other Japanese amateurs – Seki, Suzuki, Niijima, Furuta, and Kizawa. Colombini in Italy also joined in with a group of colleagues, and, of course, the above-mentioned Brian Manning in England.

As Rob McNaught will tell you, the best place to search for main belt asteroids is at or just east of the solar opposition point. As one moves away from the ecliptic, the number of minor planets drops off quickly. This is shown in Figure F18, a histogram of the declinations of all 123 known asteroids that were within 10° of the opposition longitude on September 21, 1991. From this diagram you see that although most asteroids will be within a few degrees of the ecliptic (here at Dec. = 0°), there are some to be found over 30° away – and these are the more interesting ones, of course. For this reason Marsden continually urges hunters to search well away from the ecliptic.

At opposition a main belt asteroid near the ecliptic will be moving from east to west (retrograde motion) at a rate of about 0.2° per day, or 30 seconds of arc per hour. How faint does one have to search? Figure F19 displays the predicted opposition magnitude of a random selection of main belt asteroids. Notice that since the mid-2000s, there has been little change, and in fact the eight most recently added – up to 4265 in 1990 – have an average predicted visual magnitude of 16.4. If your telescope/camera can reach to, say, magnitude 17, you are in business.

Unusual asteroids

In Chapter 4 considerable space was devoted to the 'triple-A' minor planets, the Atens, Apollos, and Amors. Because of the small sizes of these Earth-crossers, as they are sometimes called, finding one nearly always requires an ability to go faint, even though they can come close to the Earth at times and become reasonably bright. Therefore, this section of Appendix F will be brief and just note the several points worth emphasising.

Table F6 lists the Apollos and Atens discovered during the 1980s together with the more interesting orbital elements, namely a, e, i, and q. Also listed, when known, are the brightest magnitudes reached at the discovery passage; these are shown in the histogram of Figure F20. Note how quickly the numbers increase with magnitude up to 16 which, as Eleanor Helin and Carolyn Shoemaker point out in their Chapter 5 contributions, is nearing the limit of their Palomar programmes. One wonders how many chunks of rock that never get brighter than 19th or 20th magnitude pass by us every year. Probably dozens. Most of these all-but-invisible Earth-threateners would probably be smaller than 100 m in diameter. The famous Tunguska object – most experts believe that it was the nucleus of a small comet – was probably about that size. And 1991 BA is estimated to be about 10 m in diameter (see Chapter 4).

Because all had to be near the Earth to be discovered, these small planets were

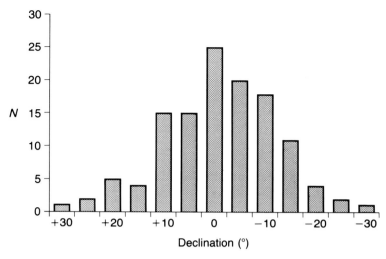

Fig. F18. The declinations of all known asteroids near opposition (± 10°) on September 21, 1991. Note that at opposition on that date the ecliptic is at Dec. = 0°.

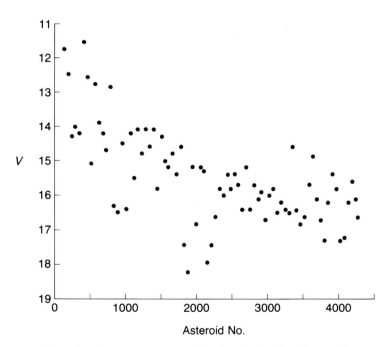

Fig. F19. The predicted opposition magnitudes of randomly selected asteroids in 1991 *vs.* asteroid number.

Table F6. Apollo–Aten asteroids discovered between January 1, 1980, and December 31, 1989

Desig.	a	e	i (°)	q	AM ("/m)	m_{pk}	ΔRA (h)	Dec. (°)	Remarks
1981 VA	2.35	0.73	21.0	0.63	6.4	16	11.6	+57	
1982 BB	1.41	0.36	21.0	0.91	1.9	16	10.3	+11	
DB	1.52	0.38	1.5	0.95	1.5	16	8.9	−18	Closest 0.028 AU
HR	1.22	0.33	2.8	0.81	3.5	17	9.4	+0	
TA	2.25	0.76	11.8	0.54	2.5	16	10.8	−0	Taurids association?
XB	1.85	0.45	3.9	1.0	11.0	14	9.4	+25	
1983 LC	1.78	0.55	1.3	0.81	14.7	16	11.1	−17	
TB	1.27	0.89	22.0	0.14	5.0	16	4.0	+59	Geminids association?
TF$_2$	2.44	0.74	15.1	0.63					
VA	2.58	0.68	16.2	0.80	3.0	16	9.7	+57	
1984 KB	2.21	0.76	4.6	0.53	7.8	13	10.3	+15	Taurids association?
QA	0.99	0.47	9.9	0.53		17			$P = 0.985$ y
1985 PA	1.41	0.30	55.6	0.99	3.2	16	10.3	−10	
1986 EB	0.97	0.28	23.4	0.70	4.0	13	10.7	+28	
JK	2.80	0.68	2.0	0.90					Closest 0.028 AU
PA	1.06	0.44	9.9	0.59	2.4	17	11.3	+2	
TO	1.00	0.51	19.8	0.49	2.5	15	9.4	−44	$P = 0.996$ y
WA	1.50	0.70	29.3	0.45	3.4	16	8.0	+13	
1987 KF	1.83	0.68	11.8	0.59	1.9	16	10.1	+17	
OA	1.47	0.58	8.9	0.61	6.3	17	10.6	+13	
QA	1.65	0.47	40.7	0.88	5.7	15	9.7	+11	
SB	2.16	0.65	3.0	0.76	1.3	16	11.3	−4	

	a	e	i	q	AM	m_{pk}	ΔRA	Dec	Delta Leonid association?
SY	1.46	0.60	5.6	0.59	1.2	17	11.5	+16	
1988 EG	1.31	0.52	3.7	0.62	9.3	13	11.7	−2	
TA	1.60	0.50	2.7	0.80	3.1	16	11.0	+12	Closest 0.009 AU
VP₄	2.26	0.65	11.7	0.79	1.2	16	11.7	+31	
XB	1.46	0.48	3.1	0.76	5.1	15	8.9	+23	
1989 AC	2.33	0.61	0.5	0.90	7.7	11	8.7	+18	
AZ	1.68	0.48	12.1	0.87	4.0	16	9.2	+21	
DA	1.68	0.41	5.2	0.99	9.5	15	9.8	+47	
FB	1.04	0.26	14.7	0.77	2.6	15	11.1	+4	
FC	1.02	0.36	5.0	0.65	90.4	12	11.7	+16	Closest 0.005 AU
JA	1.77	0.48	15.3	0.91	1.2	14	11.6	+21	
PB	1.06	0.48	8.9	0.55	5.1	13	9.7	−9	
QF	1.15	0.42	4.0	0.67	3.2	16	10.3	+1	
UP	1.87	0.48	3.9	0.98	3.6	15	9.8	+0	
UQ	0.92	0.27	1.3	0.67	3.0	14	11.4	+4	P=0.876 y
UR	1.08	0.37	10.7	0.69	3.0	18	11.1	+29	
VA	0.73	0.59	28.4	0.30	7.9	14	10.4	+27	P=0.622 y
VB	1.85	0.46	2.1	1.0	7.6	16	9.3	−3	

Notes:

a, e, i, and q are the more interesting orbital elements; AM is the angular motion at time of discovery in seconds of arc per minute; m_{pk} is the magnitude at maximumn brightness; ΔRA is the difference in right ascension between the Sun and the asteroid at time of discovery; and Dec. is the asteroid's declination at time of discovery.

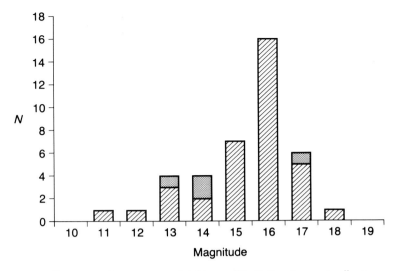

Fig. F20. The magnitude at maximum brightness of Earth-threateners: Apollos (cross-hatched) and Atens (dotted).

moving relatively rapidly across the sky when discovered. Figure F21 shows the data for the Apollos and Atens of the 1980s. Compare these speeds with that for normal asteroids near opposition, namely 0.5 minutes of arc per hour. If some night you do pick up an Apollo or an Aten, you'll almost certainly know it immediately.

Variable stars

Nearly all the information in this section comes from the IAU's *Information Bulletin on Variable Stars,* published in Budapest; the Soviet *General Catalogue of Variable Stars* from Moscow; and the journals of the oft-mentioned AAVSO and the RASNZ.

A census of all stars that are known to vary is shown in Figure F22; several comments are in order. It is important to realise that long-period Mira-type variables are the easiest to discover, and just about all the bright ones have long been found. Their magnitude range is usually large, their colours are very red, and their spectra contain conspicuous molecular absorptions. Any one of these characteristics is almost a dead giveaway.

Other pulsating stars like the Cepheids, the RR Lyraes, and the semi-regulars generally have smaller light fluctuations and more ordinary colours and spectra, but they too reveal clues to their true nature: Cepheids are extremely luminous and stand out in almost any crowd of stars; and RR Lyraes, which are found frequently in globular clusters, do their best to attract attention by varying rapidly, normally with periods of less than a day.

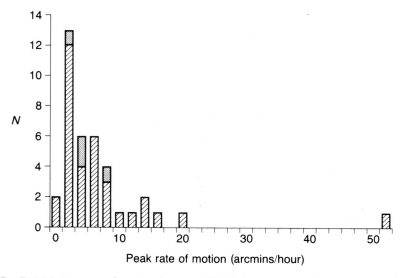

Fig. F21. Maximum angular rates of motion of Earth-threateners: Apollos (cross-hatched) and Atens (dotted).

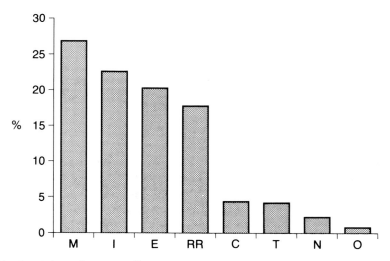

Fig. F22. Relative frequency of known variable stars in the Galaxy. M = Mira, I = semi-regular and irregular, E = eclipsing, RR = RR Lyrae, C = Cepheid, T = T Tauri, N = nova and nova-like (including SNe), O = other (mainly RV Tauri).

Table F7. *Major annual meteor showers*

Shower name	Date of maximum	Normal limits	Max. ZHR	Radiant at maximum (RA and Dec.) (h)	(°)	Comments
Quadrantids	Jan. 3	Jan. 1–Jan. 6	60	15.4	−50	Blue meteors; fine trails
Virginids	Apr. 12	Mar.–Apr.	5	14.1	−9	The two most active of several radiants
		Apr. 7–18	5	13.6	−11	in Virgo; slow long paths
Lyrids	Apr. 22	Apr. 19–Apr. 25	10	18.1	+32	Can produce fine displays
Eta Aquarids	May 5	Apr. 24–May 20	35	22.3	−1	Assoc. with Halley's Comet
Alpha Scorpids	Apr. 28	Apr. 20–	5	16.5	−24	Several weak radiants, April–July
	May 13	May 19		16.1	−24	
Ophiuchids	Jun. 10	May 19–	5	17.9	−23	Several radiants
	Jun. 20	July		17.3	−20	
Alpha Cygnids	Jul. 21	July–August	5	21.1	+48	One of a number of northern summer radiants
Capricornids	Jul. 8	July–	5	20.7	−15	Bright meteors
	Jul. 15	August		21.0	−15	
	Jul. 26					
Delta Aquarids	Jul. 29	Jul. 15–	20	22.6	−17	Meteors tend to be faint
	Aug. 7	Aug. 20	10	23.1	+2	
Piscis Australids	Jul. 31	Jul. 15–Aug. 20	5	22.7	−30	May have two maxima

Shower	Max.	Limits	ZHR	R.A. (h)	Dec.	Notes
Alpha Capricornids	Aug. 2	Jul. 15–Aug. 25	5	20.6	−10	Splendid slow yellow fireballs
Iota Aquarids	Aug. 7	July–August	8	22.2 / 22.1	−15 / −6	Rich in faint meteors; good telescopically
Perseids	Aug. 12	Jul. 23–Aug. 20	75	3.1	+58	Bright, swift meteors, many leaving trains
Piscids	Sep. 9 / Sep. 21 / Oct. 13	September–October	10 / 5 / ?	0.6 / 0.4 / 1.7	+7 / 0 / +14	Radiants also in Aries
Orionids	Oct. 22	Oct. 16–Oct. 27	25	6.4	+15	Swift with fine trains; assoc. with Comet Halley
Taurids	Nov. 4	Oct. 20–Nov. 30	10	3.7 / 3.7	+14 / +22	Very slow meteors; assoc. with Comet Encke; good photographically
Leonids	Nov. 17	Nov. 15–Nov. 20	10	10.1	+22	Swift and bright, many with trains; assoc. with Comet Tempel–Tuttle
Puppids–Velids	Dec. 9 / Dec. 26	Nov. 27–January	15	9.0 / 9.3	−48 / −65	Two of several radiants extending into Carina
Geminids	Dec. 14	Dec. 7–Dec. 16	75	7.5	+32	Bright, moderately slow; few trains; good photographically
Ursids	Dec. 23	Dec. 17–Dec. 25	5	14.5	+78	Was better in 1945 and 1986

Notes:

Exact dates will depend on when last leap year occurred. ZHR is the expected hourly rate of meteors if the radiant were in the zenith. The expected observed hourly rate, OHR = ZHR sin a, where a is the radiant elevation. The radiation will move about one degree of ecliptic longitude per day. Adapted from the *Handbook* of the British Astronomical Association.

Appendix G
Atlases and catalogues

Initial remarks

Listed below are some of the indispensable, almost indispensable, and otherwise highly informative guides to what's what in the sky. For the most part, the data given in these works are more or less permanent and unchanging, but as new objects get discovered, many catalogues must be updated. For example, Marsden's *Catalogue of Cometary Orbits* tabulates the orbital elements of all known comets up to the date of the edition publication. Information on predictions of location, brightness, etc., will be found in the journals and handbooks listed in Appendix H.

Many catalogues are available in computer format on diskettes for personal computers, through computer bulletin boards, or they can be read by one electronic means or another.

The atlases and catalogues are divided into six categories: stars (in general), comets and non-cometary nebulous objects, novae, supernovae, asteroids, and variable stars. The name of the work is given in *italics*, the author(s) in Roman letters, and the publisher in parentheses. Addresses for some of the main publishers can be found in Appendix H and addresses for clubs and societies in Appendix I. The date of publication is sometimes only approximate because of various editions or volumes.

Stars

Uranometria 2000.0. 1988 (2 volumes). Tirion, Rappaport, and Lovi (Willmann-Bell). Indispensable. The best available star atlas to magnitude (approximate) 9.5. Updates, improves upon, and replaces, the charts of the *SAO Catalogue*. Shows positions of many nebulae, galaxies, clusters, radio and X-ray sources, and quasars. Its bibliography lists many catalogues and references for these objects.

Sky Atlas 2000.0. Tirion (Cambridge University Press). Excellent maps to 8th magnitude with about 2500 extended objects included.

Sky Catalogue 2000.0 (2 volumes). 1982. Hirshfield and Sinnott (Sky Publishing Corp.). Serves as a useful companion catalogue to *Uranometria 2000.0* and especially *Sky Atlas 2000.0*. Vol. 1 lists stars to magnitude 8.0; Vol. 2 tabulates variable stars, double stars, and non-stellar objects. Diskette versions available.

The Bright Star Catalogue. 1982. Hoffleit (Yale University Observatory). A professional work giving virtually all reasonably important data on all stars brighter than approximately magnitude 6.5 and their fainter companions. Tirion's *Bright Star Atlas* maps these stars plus Messier objects.

True Visual Magnitude Photographic Star Atlas (3 volumes). 1978. Papadopoulos (Pergamon Press). 'TVMPSA' shows stars to 14th magnitude with plastic overlays to measure positions to better than ± one minute of arc, and can be used to derive approximate visual (V) magnitudes of stars to better than ±0.5 magnitudes.

Atlas Stellarum (2 volumes). 1987. Vehrenberg (Treugesell-Verlag KG). Similar to TVMPSA (see above) but incorporates blue-sensitive photographic plates to at least magnitude 14. His *Photographic Star Atlas* goes to magnitude 13 and is at half the scale.

Smithsonian Astrophysical Observatory Star Catalog (4 volumes). 1971. (Smithsonian Publ.). The most convenient source of accurate star positions (1950 equinox and epoch) — 258 997 stars to approximately magnitude 9.5. However, other more up-to-date catalogues do exist — AGK, Perth, etc. Also includes starcharts.

Hubble Space Telescope Guide Star Catalog (on two CD-ROM disks, available from the Astronomical Society of the Pacific). A catalogue of data on more than *eighteen million* celestial objects fainter than 6th magnitude. For $50, a bargain.

A Visual Atlas of the Large Magellanic Cloud. 1983. Morel (privately published, write to Mati Morel, 18 Elizabeth Cook Drive, Rankin Park, NSW, Australia). Charts down to fainter than 13th magnitude. Includes most known nebulae, variable stars, etc. Indispensable for LMC watchers.

A Visual Atlas of the Small Magellanic Cloud. 1990. Morel (privately published, see above.) The same as the above item except for the SMC.

Notes: There now exist a number of 'deep sky surveys' reaching to fainter than 20th magnitude and exposed in various spectral bands. They are all very expensive but indispensable for faint objects. Most major observatories have copies of two or more of these surveys. Some are available on CD-ROM disks and on microfiche.

The 'Centre de donnees astronomiques de Strasbourg', the CDS, in France has an immense database of stars. You specify the coordinates of your field centre, the radius of the field, and the magnitude range of stars. CDS responds with the corresponding list of stars giving virtually all known information including accurate positions, magnitudes, and colours. Write to The Librarian, Smithsonian Astrophysical Observatory, 60 Garden Street, Cambridge, MA 02138, USA, for further information.

Comets and nebulous objects

Catalogue of Cometary Orbits. Frequent editions. Marsden (Minor Planet Center, Smithsonian Astrophysical Observatory). The definitive listing of orbital elements of all known comets.

NGC 2000.0. 1988. Sinnott (Sky Publishing Corp.). A modern version of Dreyer's *New General Catalogue and Index Catalogue of Nebulae and Star Clusters*. Includes positions and descriptions of more than 13 000 objects. Available on diskette for both IBM-PC and Macs.

Observing Handbook and Catalogue of Deep-Sky Objects. 1990. Luginbuhl and Skiff (Cambridge University Press). Includes descriptions of 2050 non-stellar objects. Has faint magnitude standards!

Revised Shapley–Ames Catalogue of Bright Galaxies. 1981. Sandage and Tammann (Carnegie). Lists information on nearly all galaxies brighter than visual magnitude 12.5.

Catalogue of Galactic Planetary Nebulae. 1967. Perek and Kohoutek (Czech. Acad. of Sciences). Ancient but still the best and certainly most convenient. More than a thousand listed. With finding charts.

Note: The photographic star atlases listed in the first section show many nebulous objects fainter than 12th or 13th magnitude and thus not listed in these catalogues.

Novae

A Reference Catalogue and Atlas of Galactic Novae. 1987. Duerbeck (Reidel Publishing Co.). Gives much information on all known and many suspected galactic novae. The photographic maps are from deep sky surveys and useful mainly for identifying the nova object in its quiescent state. See additional atlases and catalogues under 'Variable Stars' below.

The Galactic Novae. 1964. Payne-Gaposchkin (Dover). A classic, much out of date, but inexpensive. Lists and describes all known galactic novae up to the date of publication.

Supernovae

The Supernova Search Charts and Handbook. 1989. Thompson and Bryan (Cambridge University Press). An excellent set of maps of the larger, brighter, and nearer galaxies with accurate visual magnitudes for a handful of stars on most charts. Especially made for the SN searcher, the charts show the galaxies much as they appear to the eye in a modest sized (20–40 cm) telescope. Flawed by a few missing stars which could lead to erroneous SN discovery reports, but still indispensable for the SN seeker. Includes useful tips on SN seeking.

SN (3 volumes). 1990. Lopez Alvarez (Editorial Acrux). Sets of photographs to about 15th magnitude of the larger, brighter, and nearer galaxies. No comparison star magnitudes are given, but they can be estimated to ±0.5 magnitudes. Makes a good companion to *The Supernova Search Charts* listed above. Inexpensive.

Asteroids

Asteroids II. 1990. (University of Arizona). Many scientific articles but Part VI contains many tables of minor-planet characteristics – discovery circumstances, family membership, magnitudes, diameters, taxonomic classifications, light-curve parameters, and pole positions. See also *Ephemerides of Minor Planets* in Appendix H.

Variable stars

General Catalogue of Variable Stars (5 volumes). Fourth edition now being published. (Nauka Publishing House, Moscow). Contains information on nearly 30 000 objects known as of 1982. This is the official listing. The first three volumes give stars alphabetically by constellation, Andromeda to Vulpecula. Volume IV contains extra-galactic data: variable stars in other galaxies plus variable galaxies. The last volume lists stars by right ascension. Does not include the many variable stars found in globular clusters (see next entry). Indispensable for the variable star hunter. Also, their *New Catalogue of Suspected Variable Stars* (1982).

Catalogue of Variable Stars in Globular Clusters (frequently issued). Sawyer Hogg (University of Toronto). Complements the above.

AAVSO Variable Star Atlas. 1982. (AAVSO). Similar to the SAO Catalog star charts but labels all variable stars that reach 9th magnitude or brighter. Also many comparison star magnitudes. Nearly indispensable and inexpensive.

Electronic catalogues

While we are still in our electronic infancy as far as everyday communication goes, professional astronomers have taken great strides in making data available directly on your home or office computer. Amateurs of course may use these services and find out immediately if a newly discovered object was or was not known before. Because this field is changing rapidly, I list here only addresses. You should write for further information.

International Stellar Data Centre, Observatoire de Strasbourg, 11 Rue de l'Universite, F67000 Strasbourg, France, Electronic address: EARN:u01117@frccsc21. As the title says, this is THE centre for all types of stellar information – coordinates, magnitudes, colours, spectral classes, variability characteristics.

Harvard–Smithsonian Center for Astrophysics, c/o The Librarian, 60 Garden Street, Cambridge, MA 02138, USA. Electronic addresses: BITNET:*username*@cfa, Internet:*username*@cfa.harvard.edu. A good place to start. Librarian Joyce M. Watson can give you all the necessary details without hassle.

Goddard Space Flight Center, c/o Dr Jaylee Burlee, Greenbelt, MD 20771, USA. TWX: 71-828-9716. This branch of NASA works closely with the Strasbourg centre and with the newly formed European Space Information System. Tell Jaylee I sent you.

picoSCIENCE, 41512 Chadbourne Drive, Fremont, CA 94539, USA. They sell IBM compatible programs with many of the star and asteroid catalogues on diskette.

Appendix H
Useful books, handbooks, and journals

Initial remarks

In the last few decades, there has been a virtual flood of astronomical books, journals, and newsletters, some more useful and informative than others. In this appendix you will find the names of those that I suggest you look into. These and many others can be found in good local libraries. Also, see the listing of clubs and societies in Appendix I; most have their own newsletters or journals.

You will find only a few non-English entries. This is not because books and journals in other languages are poor; it is partly the fault of my ignorance and partly the decision to stick to one language throughout this *Guide*.

The entries here have been divided into the three categories appearing in the title of this appendix, starting with handbooks and ending with books.

Almanacs and handbooks

The following are all issued annually. Total number of pages (in parentheses) is for 1991.

The Astronomical Almanac (556 pages). Published yearly by the Nautical Almanac Offices of the US Naval Observatory, Washington, DC and the Royal Greenwich Observatory, London. Contains precise positions of Sun, Moon, and planets; times of sunrise and sunset, circumstances of eclipses, and much more.

The Floppy Almanac. A program for MS-DOS based personal computers which reproduces much of the data contained in the major sections of the *Astronomical Almanac* and others. Comes on a single $5\frac{1}{4}$-inch DSDD diskette. Convenient and easy to use. By Tim S. Carroll, US Naval Observatory.

The Handbook of the British Astronomical Association (122 pages). Burlington House, Piccadilly, London, W1V 9AG, UK. One of the two best annual handbooks published. Some sections and articles are not repeated every year but can be ordered separately (although some are now out of print). Primarily for the UK, Australia, and New Zealand.

Observer's Handbook (236 pages). The Royal Astronomical Society of Canada, 136 Dupont Street, Toronto, Ontario M5R 1V2, Canada. One of the two best annual handbooks published. Primarily for North America. Contains more explanatory text than the BAA *Handbook*.

Solar System Ephemeris (91 pages). ALPO (see Appendix I). Data tables and charts for almost everything predictable in the solar system.

The Comet Handbook (60 pages). Smithsonian Astrophysical Observatory, 60 Garden Street, Cambridge, MA 02138, USA. A collection of orbital elements and ephemerides for most of the comets expected to become brighter than 18th or 19th magnitude during the stated year. Forty-eight comets are included in the 1991 *Handbook*.

Ephemerides of Minor Planets (448 pages). Nauka Publishing House, St Petersburg (available through White Nights Trading Co., 520 NE 83rd St, Seattle, WA, 98115, USA). This IAU-sponsored annual contains the ephemerides of all known asteroids coming to opposition in the stated year (most brighter than 16th magnitude). Also gives orbital elements and concise information on light variations.

Magazines and journals

The annual frequencies of the following appear in parentheses.

Sky & Telescope (12). PO Box 9111, Belmont, MA 02178-9111, USA. The best all-round monthly magazine published. Covers all fields of astronomy, contains many advertisements, high-quality printing.

Astronomy (12). Kalmbach Publishing Co., 21027 Crossroads Circle, PO Box 1612, Waukesha, WI 53187, USA. Somewhat less technical than *Sky & Telescope*, but very similar in format, quality, and quantity of advertisements.

Astronomy Now (12). Intra House, 193 Uxbridge Road, London W12 9RA, UK. More than a British version of *Sky & Telescope*, this journal has well-written articles, some on unusual aspects of oft-discussed topics. Advertisements of companies in the UK.

Journal of the British Astronomical Association (6). The British Astronomical Association, Burlington House, Piccadilly, London, W1V 9AG, UK. Interesting articles and refreshingly little advertising. The best of the club-and-society journals.

Mercury (6). Astronomical Society of the Pacific, 390 Ashton Avenue, San Francisco, CA 94112, USA. A non-technical journal with interesting articles on current research and history of astronomy. Good annual listings of recommended books, computer programs, and other resources.

The International Comet Quarterly (5). Smithsonian Astrophysical Observatory, 60 Garden Street, Cambridge, MA 02138, USA. 'A journal devoted to news and observations of comets'. One issue each year is the *Comet Handbook* described above.

The Astronomer (12). 16 Westminster Close, Kempshott Rise, Basingstoke, Hants, RG22 4PP, UK. About 'comets, asteroids, photo notes, deep sky, and [contains] general articles'. A good journal for the observing amateur.

The Minor Planet Bulletin (4). 10385 East Observatory Drive, Tucson, AZ 85747, USA. The journal of the Minor Planets Section of ALPO (see Appendix I). Publishes information on astrometry, photometry, and special asteroid events. Index of scientific papers and news notes are especially informative.

Minor Planet Observer (12). Bdw Publishing, Box 818, Florissant, CO 80816, USA. An enjoyable monthly devoted to asteroids with maps showing where to see them (generally brighter than mag. 13.5). In addition *MPO* publishes predictions of asteroid–star occultations plus occasional articles. Editor Brian Warner also markets asteroid software (orbits and ephemerides).

International Astronomical Union (IAU) Circulars (many). Smithsonian Astrophysical Observatory, 60 Garden Street, Cambridge, MA 02138, USA. The official announcements of discoveries and other information on transient astronomical phenomena of all sorts.

Comet Comments (12 +). Don Machholz (PO Box 1716, Colfax, CA 95713, USA) prints a newsletter that comes out at least once a month. Often describes how new discoveries were made.

Comet Circulars (irregular). J. E. Bortle (R.R. No.1 Box 198, Gold Road, Stormville, NY 12582, USA) also sends out brief newsletters on new comets as they get discovered.

Information Bulletin on Variable Stars (many). Konkoly Observatory, Budapest, Hungary (almost always in English). The *IBVS* publishes short articles mainly about observations and sometimes rather technical.

Note: The European magazines best known to me are *Sterne und Weltraum* in Germany, *Le Ciel* in France, and *Orion* in Switzerland. All are excellent. In many other countries there are popular magazines of high quality. Just be sure they are concerned with astronomy, not astrology.

Books

Many of the best books on the market soon go out of date in this rapidly changing field of astronomy, and so, in addition to the several titles and authors listed below, I provide you with the names of a few leading publishers and book dealers. Reliable book reviews can be found in many of the magazines and journals listed above. Special mention should be made of the Astronomical Society of the Pacific's 'Astronomy Books of [the Year]' which appears in the magazine *Mercury*, usually in the May–June issue. The books below that are prefixed with an asterisk received one of the ASP's Book of the Year awards. Many of the mini-reviews given here were extracted from Editor Andrew Fraknoi's frank comments therein.

General astronomy

Every sky searcher should have near at hand a good, standard college-level textbook on astronomy. Of the many that are used in universities in the USA, the ones listed below I know and can recommend. All have catchy titles to attract the poets and polysci majors, and one or two can probably be found in your local bookstore (usually the one used in the nearest university). Here I simply list them alphabetically by author with the publisher in parentheses. Because new editions are constantly appearing, no dates are given – but make sure what you buy is recent.

Abell, Morrison, and Wolf (Saunders), Chaisson (Prentice-Hall), DeVorkin (Smithsonian), Goldsmith (Benjamin/Cummings), Hartmann (Wadsworth), Jastrow and Thompson (John Wiley), Kaufmann (Freeman), Kutner (Harper and Row), Parker, B. (Harcourt, Brace, Jovanovich), Pasachoff (Saunders), Seeds (Wadsworth), Snow (West), Zeilik and Smith (Saunders).

Comets

Halley's Comet generated a rash of books, many of them very good, including some listed here.

The Comet Book. Chapman, R. and Brandt, J. (1984, Jones and Bartlett). Inspired by Halley's but covers much more. Filled with authoritative, up-to-date information.

Observe: Comets. Edberg, S. and Levy, D. (1985, Astronomical League, 4 Klopfer St, Pittsburgh, PA 15209, USA). Guide for serious amateurs. Background material and techniques for those who want to do extensive comet hunting, observing, and photography.

A Decade of Comets. Machholz, D. (1985, privately published). Highly useful information for the comet hunter.

**Comet.* Sagan, C. and Druyan, A. (1985, Random House). Stylishly written, well-organised with excellent illustrations.

**The Mystery of Comets.* Whipple, F. (1985, Smithsonian). A non-technical introduction to comets. A rare combination of scientific autobiography and popularisation, this book belongs in every comet-fancier's library.

The Origin of Comets. Bailey, M. E., Clube, S. V. M., and Napier, W. M. (1990, Pergamon). Combines ancient history with modern theories.

Novae

Duerbeck's *A Reference Catalogue and Atlas of Galactic Novae*, listed in Appendix G, contains information on individual novae. Alas, there has been no recent general book devoted to the subject. Still a classic, however, is the following:

The Galactic Novae. Payne-Gaposchkin, C. (1964, Dover). Much out of date, but inexpensive. Lists and describes all known galactic novae up to the date of publication.

Supernovae

Several books have been published recently dealing primarily with SN 1987A in the LMC. Also, the *Handbook* that accompanies the Thompson–Bryan *Supernova Search Charts* (see Appendix G) has some excellent sections. The two books listed here are more general.

*The Supernova Story. Marschall, L. (1988, Plenum). Superbly written book. The introduction of choice for anyone who wants to learn more about supernovae.
Supernovae. Murdin, P. and L. (1985, Cambridge University Press). A good, mostly non-technical introduction.

Minor planets

Asteroids II (listed in Appendix G) is for the most part highly technical. The following two books are more readable.

*Asteroids: Their Nature and Utilization'. Kowal, C. (1988, Ellis Horwood/John Wiley). A fascinating little book by an asteroid expert.
Introduction to Asteroids. Cunningham, C. (1988, Willmann-Bell). Presents the latest scientific findings.

Variable stars – and stars in general

The *AAVSO Variable Star Atlas* (see Appendix G) comes with a useful book on how to observe variables (with an article by J. E. Bortle on how to search for comets). More information on variables and stars in general can be found in:

Observing Variable Stars. Levy, D. (1989, Cambridge University Press). A good introduction to all types of variable stars and how to observe them.

The Classification of Stars. Jaschek, C. and M. (1990, Cambridge University Press). Handbook on the tools, methods, and results of stellar classification.

Stars and their Spectra. Kaler, J. (1990, Cambridge University Press). A good introduction to stellar spectra by an expert.

Meteors and meteorites

There is much overlap in this category since all meteorites were once meteors – and before that meteoroids (i.e., small asteroids).

Meteor Showers. Kronk, G. W. (1988, Enslow). A descriptive catalogue. A month-by-month listing of most known or suspected meteor showers with their histories and descriptions.

Thunderstones and Shooting Stars: The Meaning of Meteorites. Dodd, R. (1986, Harvard). A clear introduction to the recovery, classification, and study of meteorites.

Meteorites and their Parent Planets. McSween, H. (1987, Cambridge University Press). An authoritative introduction to meteorites and how and where they were formed.

The solar system, orbits, computing

Thousands of books are available on the planets; the two listed are recent entries, and one, *The New Solar System*, already something of a classic. There are also thousands of texts on computing; Duffet-Smith's covers much of the more useful astronomical territory in BASIC. Not listed are some of the dozens of astronomical computer programs now available and increasing rapidly in number. The ASP's annual *Astronomy Catalog* lists and reviews most of the better ones.

The New Solar System. Beatty, J. K. and Chaikin, A. (1990, Cambridge/Sky Publishing Corp.). Although popular in style, it has much scientific value.

Fundamentals of Celestial Mechanics. Danby, J. M. A. (1988, Willmann-Bell). Basic orbit theory and related problems. A classic; requires college mathematics.

Astronomy With Your Personal Computer. Duffet-Smith, P. (1985, Cambridge University Press). A collection of 26 astronomical subroutines (written in BASIC) which can be used separately or linked together into more complex programs.

Practical Astronomy With Your Calculator. Duffet-Smith, P. (1989, Cambridge University Press). A 'must' if you have no PC.

Introduction to Basic Astronomy with a PC Computer. Lawrence, J. L. (1989, Willmann-Bell). A practical introduction to astronomical computing with a floppy diskette included.

Photography, observational methods

For the best information on CCDs and their use, see recent issues of *Sky & Telescope* and *Astronomy*. This field is changing so rapidly that much of any book is out of date before it gets printed.

The Cambridge Astronomy Guide. Liller, B. and Mayer, B. (1990, Cambridge University Press). The 'Bill and Ben Book'. The theory and practice of making observations and photographs of the sky. Read Ben Mayer's 'odd' chapters if you want to be inspired. My even-numbered chapters try to complement his text.

Burnham's Celestial Handbook: An Observer's Guide to the Universe Beyond the Solar System, 3 volumes. Burnham, R. (1978, Dover). A comprehensive coverage to thousands of celestial objects.

Webb Society Deep-Sky Observer's Handbooks. 8 volumes (available through Willmann-Bell). A series of observer's manuals, written by experts. The volume titles are: Double Stars, Planetary and Gaseous Nebulae, Open and Globular Clusters, Galaxies, Clusters of Galaxies, Anonymous Galaxies, The Southern Hemisphere, Variable Stars.

Astrophotography. Gordon, B. (1985, Willmann-Bell). A thorough guide to astronomical photography. Could be used as a textbook.

A Manual of Advanced Celestial Photography. Wallis, B. and Provin, R. (1988, Cambridge). This technical handbook provides a wealth of information about techniques and instruments.

Miscellaneous

Stargazers. Dunlop, S. and Gerbaldi, M. (1988, Springer-Verlag). Articles and summaries of articles from an IAU Colloquium 'The Contribution of Amateurs to Astronomy'. See especially Marsden's article.

The Sky: A User's Guide. Levy, D. (1991, Cambridge University Press). An enjoyable book on the variety of objects in the night sky.

Dealers and publishers

Cambridge University Press. The Edinburgh Building, Cambridge CB2 2RU, UK. The world's largest publisher of astronomical books of all types. Many interesting titles not listed above. Get their catalogue! Watch their ads!

Willmann-Bell, Inc. PO Box 35025, Richmond, VA 23235, USA. One of the largest suppliers of astronomical catalogues and literature. Also has an extensive collection of computer software. Excellent service. Write for their catalogue.

Sky Publishing Corporation. PO Box 9111, Belmont, MA 02178-9111, USA. Sell many useful books on astronomy besides publishing many themselves. See their advertisements in *Sky & Telescope.*

Herbert A. Luft. 46 Woodcrest Drive, Scotia, NY 12302, USA. Stocks hundreds of astronomical and scientific books – and searches for others if you ask him.

The following companies publish many professional-level books on astronomy, especially those containing articles presented at symposia and colloquia. Some are very much at the level of the advanced amateur. You should write for their catalogues.

Springer-Verlag, 175 Fifth Avenue, New York, NY 10010, USA.

D. Reidel Publishing Company, PO Box 17, 3300 AA Dordrecht, The Netherlands.

Kluwer Academic Publishers, 101 Philip Drive, Norwell, MA 02061, USA, and PO Box 560, London N11 2UX, UK.

Appendix I
Associations, clubs, and societies

Initial remarks

In 1990 there were 59 astronomy clubs just in California alone, according to *Sky & Telescope*. Every year *S&T* lists, to the best of its knowledge, all the amateur clubs, planetariums, and observatories in the USA and Canada (September issues). In addition *S&T* publishes telephone 'hotlines', computer bulletin boards, and 'Special Interest Organizations', a number of which we list below.

Annually, Heck and Manfroid publish the 'International Directory of Amateur Astronomical Societies'. For a copy write to A. Heck at Observatoire Astronomique, 11 rue de l'Universite, F-67000, Strasbourg, France.

The organisations listed here are the longer-established and better known ones (to me, at any rate), first in the USA and Canada, then internationally – but English-speaking.

USA and Canada

American Association of Variable Star Observers, Janet Mattei, 25 Birch Street, Cambridge, MA 02138. Publication: *Journal*.

American Meteor Society, David Meisel, Dept. of Physics–Astronomy, SUNY-Geneseo, Geneseo, NY 14454.

Association of Lunar and Planetary Observers, Harry D. Jamieson, PO Box 143, Heber Springs, AR 72543. Also Box 16131, San Francisco, CA 94116. Several publications.

Astronomical League, Merry Edenton-Wooten, 6235 Omie Circle, Pensacola, FL 32504.

Astronomical Society of the Pacific, Andrew Fraknoi, 390 Ashton Avenue, San Francisco, CA 94112. Publications: *Mercury, Publications*.

International Meteor Organization, Peter Brown, North American Secretary, 181 Sifton Avenue, Fort McMurray, Alta T9H 4V7, Canada.

Meteoritical Society, H. Y. McSween, Geological Sciences Division, University of Tennessee, Knoxville, TN 37996.

PROBLICOM Sky Survey, Ben Mayer, 1940 Cotner Avenue, Los Angeles, CA 90025. Publication: *Newsletter*.

Royal Astronomical Society of Canada, 136 Dupont Street, Toronto, Ontario M5R 1V2, Canada. Publications: *Journal, Newsletter, Observer's Handbook.*

Sunsearch–Supernova Search, Steve Lucas, 14400 S Kolin Avenue, Midlothian, IL 60445. Publication: *Newsletter.*

Other English-speaking clubs and societies

British Astronomical Association, Burlington House, Piccadilly, London W1V 9AG, UK. Publications: *Bulletin, Newsletter, Handbook.*

Society of Meteoritophiles, P. M. Bagnall, 9 Airedale, Hadrian Lodge W, Wallsend, Tyne and Wear NE28 8TL, UK.

Royal Astronomical Society of New Zealand, Astronomical Research Limited, PO Box 3093, Greerton, Tauranga, NZ. Publication: *Publications* of the Variable Stars Section.

National Association of Planetary Observers, Geoff McNamara, PO Box 2, Riverwood, NSW 2210, Australia. Publication: *Iris.*

Appendix J
Where to buy it

Initial remarks

One look at the pages of *Sky & Telescope* or *Astronomy* or *Astronomy Now* should convince you that astro-marketing is a fierce, if not a big business. You can also guess that many companies do not survive long in this competitive world of selling amateur astronomy; the market is limited.

I feel that for a *Guide* like this one, a ready reference of places where equipment can be bought should include companies that (1) have the largest inventories, (2) carry items that are unique or difficult to find elsewhere, (3) have shown that their service is friendly, accurate, and fast, (4) will accept credit card mail orders and, of course, (5) are well known for the quality of their products.

In the first list I have included (alphabetically) the dealers and manufacturers that I am familiar with, either directly or indirectly, and suggest you consider these companies when in the market. All are located in the USA, but there should be no problem ordering from other countries – except for the duty that you will have to pay.

In the following shorter lists are the companies in Great Britain and Australia whose advertising *suggests* to me that they meet the criteria given in the preceding paragraph. I list them so that you can write for further information or for their catalogues.

A number of the companies were mentioned specifically by me or by the contributors to Chapter 5, and while that strongly implies that their products are good, it does not necessarily mean that they are the best. And if I have omitted companies that you think should be included, please let me know.

Companies in the USA

ADORAMA. 42 West 18th Street, New York, NY 10011. 'The photography people' but with a large stock of name brands of telescopes and accessories. My experience has been that the lowest prices for cameras can be found not in airport 'Duty Free' shops but in downtown New York City where a number of stores buy duty free in large quantities – and do not have to pay the high rents charged in airports. See advertisements in, for example, *Popular Photography*, said, by my professional photographer-daughter, to be the best of the slick 'fotomags'.

ASTRO LINK. PO Box 1978, Spring Valley, CA 92077. One of the leading suppliers of CCD equipment.

EDWARD R. BYERS, CO. 29001 West Highway 58, Barstow, CA 92311. 'The best is always a bargain' is their motto, and Byers has long been one of the very best manufacturers of worm gears and telescope drives.

CELESTRON INTERNATIONAL. PO Box 3578, 2835 Columbia Street, Torrance, CA 90503. With Meade, the largest manufacturer of popular Schmidt–Cassegrain telescopes, mounts, and accessories. Also manufactures Schmidt cameras in at least two sizes, 8- and 14-inch.

COULTER OPTICAL. PO Box K, Idyllwild, CA 92349-1107. Mirrors plus complete Newtonian systems with Dobsonian mounts.

EDMUND SCIENTIFIC CO. 101 E Gloucester Pike, Barrington, NJ 08007-1380. The largest supplier of all sorts of things – lenses, mirrors, filters, prisms, gratings, as well as binoculars, telescopes, microscopes, and laboratory equipment.

LAZERSON, Dr H. E. 8540 S Sepulueda Blvd, Suite 115, Los Angeles, CA 90045. Blink comparator.

LUMICON. 2111 Research Drive, Livermore, CA 94550. 'The world center for astrophotography' – and much more. Besides all the well-known astro-products that you can think of, they sell a number of remarkable filters (deep-sky, comet, H-alpha, H-beta, Oxygen-III, ultra-high contrast), and a complete 'film-hypersensitization kit'. Owner Dr Jack Marling is both a trained scientist and a serious amateur astronomer.

MEADE INSTRUMENTS. 1675 Toronto Way, Costa Mesa, CA 92626. With Celestron, the largest manufacturer of popular Schmidt–Cassegrain telescopes, mounts, and accessories. Battles continue as to which is better: Celestron or Meade. Write for their catalogue, but you must buy elsewhere.

ORION TELESCOPE CENTER. 2450 17th Avenue, PO Box 1158-S, Santa Cruz, CA 95061. A large supply of eyepieces, finder scopes, focussers, filters.

PARKS OPTICAL. 270 Easy Street, Simi Valley, CA 93065. One of the best manufacturers of Newtonian reflectors, but also sells other types of telescopes and mounts.

PHOTOMETRICS. 3440 E Britannia Drive, Tucson, AZ 85706. One of the leading suppliers of CCD equipment.

SANTA BARBARA INSTRUMENT GROUP (SBIG). 1482 East Valley Road, Suite #601, Santa Barbara, CA 93108. One of the leading suppliers of CCD equipment.

SCOPE CITY. 679 Easy Street, Simi Valley, CA 93065. 'A huge selection of telescopes, binoculars . . .', etc. Many outlets, including one in Canada.

SPECTRASOURCE INSTRUMENTS. PO Box 1045, Agoura Hills, CA 91376. One of the leading suppliers of CCD equipment.

TELE VUE OPTICS. 20 Dexter Plaza, Pearl River, NY 10965. Rich field refractors, mounts, and accessories, including Plössl eyepieces.

TELRAD. Steve Kufeld, 7092 Betty Drive, Huntington Beach, CA 92647. Manufactures the zero-power reflex finder used by many amateurs.

ROGER TUTHILL, INC. Box 1086-A, Mountainside, NJ 07092. Has an interesting selection of accessories.

UNITRON, INC. 170 Wilbur Place, PO Box 469, Bohemia, NY 11716. Refractor telescopes and accessories 'designed to the highest optical standards'.

VERNONSCOPE and CO. Candor, NY 13743. Manufactures compact refractors.

WILLMANN-BELL, INC. PO Box 35025, Richmond, VA 23235. Besides books and computer programs, sells mirror-making kits and supplies.

Companies in Great Britain

ASTRO SYSTEMS LTD. 1A Hartley Road, Luton, Beds LU2 0HX. Newtonian telescopes and mounts, binoculars.

BROADHURST CLARKSON AND FULLER. 63 Farringdon Road, London EC1M 3JB. Since 1785, major dealer in telescopes and binoculars of many brands.

CARL LINGARD. 89 Falcon Crescent, Clifton, Swinton, Manchester M27 2JP. A wide selection of telescopes, mirror kits, optics, and accessories.

CRABB OPTICAL UK. 543 Old Chester Road, Birkenhead. Ready-made mirrors and mirror sets for assembling your own telescope.

DALSERF OPTICS. The Old School House, Manse Brae, Dalserf, Lanarkshire ML9 3BN. Newtonians and mounts, ex-government lenses and prisms, telescopes made to order.

OPTICRAFT LTD. Unit 4, Queen Street Mill, Harle Syke, Burnley. 8-inch and 12-inch Maksutov telescopes with mounts.

ORION OPTICS. Unit 3M, ZAN Industrial Park, Crewe Road, Wheelock, Sandbach, Cheshire CW11 0QD. A wide selection of telescopes, optics, and accessories.

OSBORNES OPTICS. 139 Dean House, Eastfield Avenue, Walker, Newcastle upon Tyne NE6 4UU. Authorised dealer for Tele Vue (see under USA manufacturers).

'SCOPE CITY. 71 Bold Street, Liverpool. Telescopes, binoculars, books, and accessories.

WISE INSTRUMENTS LTD. Unit 9, Hollins Business Centre, Marsh Street, Stafford ST16 3BG. 'Specialist optics for astronomy and industry'.

Companies in Australia

ASTRO OPTICAL. 53 Hume Street, Crows Nest NSW 2065, and Mid City Arcade, 200 Bourke Street, Melbourne, VIC 3000. Manufactures Newtonians and both German and Dobsonian mounts. Stocks Celestron products.

K. M. RYAN. PO Box V105, Mt Druitt Village, NSW 2770. Supplier of many astro-films, including hypersensitisation (with optional shipping in dry ice). Stockists in Auckland (SKYLAB) and Christchurch (BLAXALL SCIENCE CO.), New Zealand.

OMNI OPTICS. Unit 10, 43 Hutton Street, Osborne Park 6017. Tele Vue optics (see under USA manufacturers).

PRECISION OPTICS AUSTRALIA. Freepost 20A, 138 Railway PDE, Leederville, WA 6007. Meade instruments dealer.

THE TELESCOPE AND BINOCULAR SHOP. 310 George Street, Sydney 2000. 'Huge range' plus Tele Vue eyepieces (see under USA manufacturers).

Name index

Subject index